29th Development of Materials Science in Research and Education (DMSRE29)

IOP Conference Series: Materials Science and Engineering
Volume 726

Nova Lesna, Slovakia
2 - 6 September 2019

Editors:

Maria Behulova
Zdenek Kozisek
Zdenek Potucek

ISBN: 978-1-7138-2585-2
ISSN: 1757-8981

Printed from e-media with permission by:

Curran Associates, Inc.
57 Morehouse Lane
Red Hook, NY 12571

Some format issues inherent in the e-media version may also appear in this print version.

Institute of Physics
Dirac House, Temple Back
Bristol BS1 6BE UK

Phone: 44 1 17 929 7481
Fax: 44 1 17 920 0979

techtracking@iop.org

Additional copies of this publication are available from:

Curran Associates, Inc.
57 Morehouse Lane
Red Hook, NY 12571 USA
Phone: 845-758-0400
Fax: 845-758-2633
Email: curran@proceedings.com
Web: www.proceedings.com

TABLE OF CONTENTS

IOP Conf. Series: Materials Science and Engineering **726** (2020) 011001 doi:10.1088/1757-899X/726/1/011001

Preface

The 29th seminar "Development of Materials Science in Research and Education" held on 2-6 September 2019 in the Slovak Republic, in the Eufória hotel located in Nová Lesná, a willage with the most beautiful view of the High Tatras.

The first Joint Seminar in these series was held at Gabčíkovo in 1991 by the initiative of the Czech and Slovak Association for Crystal Growth and the Slovak Expert Group of Solid State Chemistry and Physics. The objective of this meeting offered an opportunity to Czech and Slovak teachers and scientists as well as guests from other countries who are working in the field of Materials Science to present their recent results and experience and to exchange new ideas and information.

The scientific sessions covered the following topics on materials science:

- Trends in the development of materials research
- Education of materials science at the universities
- Information about the research programmes of individual institutions
- Information about equipment for preparation and characterization of materials
- Results of materials research

This workshop was aimed at the development of a stimulating atmosphere of cooperation, support of genuine dissemination of scientific ideas and propagation of materials science education. At the seminar, 44 participants presented 42 interesting talks.

Editors

29th Joint Seminar

Development of Materials Science in Research and Education

Hotel Eufória, Nová Lesná, 2 - 6 September 2019

Organized by

Czechoslovak Association for Crystal Growth
Slovak Expert Group of Solid State Chemistry and Physics

Under the auspices of

Slovak Society for Industrial Chemistry
Faculty of Chemical and Food Technology STU in Bratislava
Faculty of Materials Science and Technology STU in Bratislava
Czech Academy of Sciences, Institute of Physics
Regional Committee of Czech and Slovak Crystallographers
Crystallographic Society

IOP Conf. Series: Materials Science and Engineering **726** (2020) 011002 doi:10.1088/1757-899X/726/1/011002

Scientific Committee

Organizing and Program Committee

Editors

SPONSORS

Slovak Society for Industrial Chemistry Koľová 15, 815 94 Bratislava, Slovak Republic
VWR International GmbH, Plynárenská 1, 821 09 Bratislava, Slovak Republic
Carl Zeiss Slovakia, s.r.o., Račianska 12481/77/A, 831 02 Bratislava, Slovak Republic
Kvant, spol. s r.o., FFMI UK, Mlynská dolina, 842 48 Bratislava, Slovak Republic
Hotel Eufória, Nová Lesná 339, Slovak Republic

The Conference was organized with the support of the Ministry of Education, Science, Research and Sport of the Slovak Republic within the Research and Development Operational Programme for the project "University Science Park of STU Bratislava", ITMS 26240220084, co-funded by the European Regional Development Fund.

29th Joint Seminar

Development of Materials Science in Research and Education

Hotel Eufória, Nová Lesná, 2 - 6 September 2019

LIST OF PARTICIPANTS

Paulína Babincová
Slovak University of Technology in Bratislava
Faculty of Materials Science and Technology in Trnava
Ulica Jána Bottu č. 2781/25
917 24 Trnava
Slovak Republic
paulina.babincova@stuba.sk

Mária Behúlová
Slovak University of Technology in Bratislava
Faculty of Materials Science and Technology in Trnava
Ulica Jána Bottu č. 2781/25
917 24 Trnava
Slovak Republic
maria.behulova@stuba.sk

Ondřej Bílek
Tomas Bata University in Zlín
Faculty of Technology
Vavrečkova 275
760 01 Zlín
Czech Republic
bilek@utb.cz

Bohumil Bohunický
Kvant, spol. s r.o.
FFMI UK
Mlynská dolina
842 48 Bratislava
Slovak Republic
bohunicky@kvant.sk

Marek Brezovan
Slovak University of Technology in Bratislava
Faculty of Chemical and Food Technology
Radlinského 9
812 37 Bratislava
Slovak Republic
marek.brezovan@stuba.sk

Ľubomír Čaplovič
Faculty of Materials Science and Technology
Institute of Materials Science
Jána Bottu 25
917 24 Trnava
Slovak Republic
lubomir.caplovic@stuba.sk

Juraj Černák
P. J. Šafárik University in Košice
Institute of Chemistry
Moyzesova 11
041 54 Košice
Slovak Republic
juraj.cernak@upjs.sk

Ľubor Dlháň
Slovak University of Technology in Bratislava
Faculty of Chemical and Food Technology
Radlinskeho 9
812 37 Bratislava
Slovak Republic
lubor.dlhan@stuba.sk

Jozef Dobrovodský
Slovak University of Technology in Bratislava
Faculty of Materials Science and Technology in Trnava
Ulica Jána Bottu č. 2781/25
917 24 Trnava
Slovak Republic
jozef.dobrovodsky@stuba.sk

Jaromír Drápala
VSB-Technical University of Ostrava
Faculty of Materials Science and Technology
17. listopadu 15
70800 Ostrava - Poruba
Czech Republic
Jaromir.Drapala@vsb.cz

Peter Drobný
Slovak University of Technology in Bratislava
Faculty of Materials Science and Technology in Trnava
Jána Bottu 8857/25
917 24 Trnava
Slovak Republic
peter.drobny@stuba.sk

Jan Havlíček
Crytur spol. s.r.o.; UCT Prague
University of Chemistry and Technology
Technická 5
166 28 Praha 6
Czech Republic
havlicek.honza7@seznam.cz

Vladimír Jorík
Slovak University of Technology in Bratislava
Faculty of Chemical and Food Technology
Radlinského 9
812 37 Bratislava
Slovak Republic
vladimir.jorik@stuba.sk

Jana Knedlová
Tomas Bata University in Zlín
Faculty of Technology
Vavrečkova 275
760 01 Zlín
Czech Republic
knedlova@utb.cz

Marian Koman
Slovak University of Technology in Bratislava
Faculty of Chemical and Food Technology
Radlinského 9
812 37 Bratislava
Slovak Republic
marian.koman@stuba.sk

Zdeněk Kožíšek
Czech Academy of Sciences
Institute of Physics
Cukrovarnická 10
162 00 Praha 6
Czech Republic
kozisek@fzu.cz

Jan Krčil
CTU in Prague, Faculty of mechanical engineering
Department of materials engineering
Karlovo náměstí 13
121 35 Praha 2
Czech Republic
jan.krcil@fs.cvut.cz

Milena Kubišová
Tomas Bata University in Zlín, Faculty of Technology
Department of Production Engineering
Vavrečkova 275
760 01 Zlín
Czech Republic
mkubisova@utb.cz

Vladimír Kucek
Univerzita Pardubice
Faculty of Chemical Technology
Studentská 573
532 10 Pardubice
Czech Republic
vladimir.kucek@upce.cz

Vladimír Kuchtanin
Slovak University of Technology in Bratislava
Faculty of Chemical and Food Technology
Radlinského 9
812 37 Bratislava
Slovak Republic
vladimir.kuchtanin@stuba.sk

Peter Letovanec
Carl Zeiss Slovakia, s.r.o.
Research Microscopy Solutions
Račianska 12481/77/A
831 02 Bratislava
Slovak Republic
peter.letovanec@zeiss.com

Monika Losertová
VSB-Technical University of Ostrava
Faculty of Materials Science and Technology
17. listopadu 15
70800 Ostrava - Poruba
Czech Republic
mlosertova@vsb.cz

Vladimír Mára
Czech Technical University in Prague
Department of Materials Engineering
Karlovo náměstí 13
12135 Praha 2
Czech Republic
vladimir.mara@fs.cvut.cz

Lenka Markovičová
University of Žilina
Faculty of Mechanical Engineering
Univerzitná 8215/1
010 26 Žilina
Slovak Republic
lenka.markovicova@fstroj.uniza.sk

Jaroslava Maroszová
Slovak University of Technology in Bratislava
Faculty of Chemical and Food Technology
Radlinského 9
812 37 Bratislava
Slovak Republic
jaroslava.maroszova@stuba.sk

Maroš Martinkovič
Slovak University of Technology in Bratislava
Faculty of Materials Science and Technology in Trnava
Jána Bottu 8857/25
917 24 Trnava
Slovak Republic
maros.martinkovic@stuba.sk

Radoslav Milde
Tomas Bata University in Zlín
Faculty of Technology
Vavrečkova 275
760 01 Zlín
Czech Republic
rmilde@utb.cz

Stanislav Minárik
Slovak University of Technology in Bratislava
Faculty of Materials Science snd Technology in Trnava
Jána Bottu 8857/25
917 24 Trnava
Slovak Republic
stanislav.minarik@stuba.sk

Martin Necpal
Slovak University of Technology in Bratislava
Faculty of Materials Science snd Technology in Trnava
Jána Bottu č. 2781/25
917 24 Trnava
Slovak Republic
martin.necpal@stuba.sk

Blažena Papánková
Slovak University of Technology in Bratislava
Faculty of Chemical and Food Technology
Radlinského 9
812 37 Bratislava
Slovak Republic
blazena.papankova@stuba.sk

Dušan Poštulka
ON Semiconductor Czech Republic
1. máje 2634
756 61 Rožnov pod Radhoštěm
Czech Republic
dusan.postulka@onsemi.com

Zdeněk Potůček
Czech Technical University in Prague
Faculty of Nuclear Sciences and Physical Engineering
Trojanova 13
120 00 Praha
Czech Republic
potucek@fzu.cz

Miroslava Puchoňová
Slovak University of Technology in Bratislava
Faculty of Chemical and Food Technology
Radlinského 9
812 37 Bratislava
Slovak Republic
mirka.puchonova@gmail.com

Zdeněk Remeš
Czech Academy of Sciences
Institute of Physics
Na Slovance 1999/2
182 00 Praha 8
Czech Republic
remes@fzu.cz

Pavlína Ruleová
University of Pardubice
Faculty of Chemical Technology
Studentská 573
532 10 Pardubice
Czech Republic
Pavlina.Ruleova@upce.cz

Pavla Ryparová
Czech Technical University in Prague
Faculty of Civil Engineering
Thákurova 7
16629 Praha 6
Czech Republic
pavla.ryparova@fsv.cvut.cz

Martin Sahul
Slovak University of Technology in Bratislava
Faculty of Materials Science snd Technology in Trnava
Ulica Jána Bottu č. 2781/25
917 24 Trnava
Slovak Republic
martin.sahul@stuba.sk

Zuzana Štrbová
VUCHT a.s.
Nobelova 34
836 03 Bratislava
Slovak Republic
zstrbova@vucht.sk

Jan Strnad
Tomas Bata University in Zlín
Faculty of Technology
Vavrečkova 275
760 01 Zlín
Czech Republic
strnad.jan93@seznam.cz

Libuše Sýkorová
Tomas Bata University in Zlín
Faculty of Technology
Vavrečkova 275
760 01 Zlín
Czech Republic
kubisovam@seznam.cz

IOP Conf. Series: Materials Science and Engineering **726** (2020) 011003 doi:10.1088/1757-899X/726/1/011003

Štefan Vrtiel
Slovak University of Technology in Bratislava
Faculty of Materials Science and Technology in Trnava
Ulica Jána Bottu č. 2781/25
917 24 Trnava
Slovak Republic
stefan.vrtiel@stuba.sk

Milan Žaludek
Tomas Bata University in Zlín
Faculty of Technology
Vavrečkova 275
760 01 Zlín
Czech Republic
zaludek@utb.cz

Hüsnügül Yilmaz Atay
İzmir Katip Çelebi University
Faculty of Engineering and Architecture
Çiğli 35620
İzmir
Turkey
hgulyilmaz@gmail.com

Viera Zatkalíková
University of Žilina
Faculty of Mechanical Engineering
Univerzitná 8215/1
010 15 Žilina
Slovak Republic
viera.zatkalikova@fstroj.uniza.sk

IOP Conf. Series: Materials Science and Engineering **726** (2020) 011004 doi:10.1088/1757-899X/726/1/011004

Peer review statement

All papers published in this volume of *IOP Conference Series: Materials Science and Engineering* have been peer reviewed through processes administered by the proceedings Editors. Reviews were conducted by expert referees to the professional and scientific standards expected of a proceedings journal published by IOP Publishing.

Morphological changes of the PVD coatings after isothermal annealing

P Babincová, M Sahul, P Drobný and Ľ Čaplovič

Slovak University of Technology in Bratislava, Faculty of Materials Science and Technology in Trnava, Ulica Jána Bottu č. 2781/25, 917 24 Trnava, Slovak Republic

paulina.babincova@stuba.sk

Abstract. The main of this work is to investigate the impact of the isothermal annealing on the structure and morphological changes on the PVD hard coatings. Three different coatings based on the AlTiN system were deposited onto cemented carbide substrates containing 6 wt% of Co. Two coatings AlTiN/TiAlN and AlTiN/TiN with an initial thickness of 3 μm has a coarse-grained columnar structure with a chemically graded nanomultilayering consisting of alternating Ti-rich and Al-rich layers. In the case of the AlTiN/TiAlN coating, a total of 20 layers were applied. Ten layers had a thickness of 200 nm (AlTiN) and 10 layers with the thickness of 100 nm (TiAlN). The same number of nanolayers was also retained for the AlTiN/TiN multilayer coating. The TiN layers were 100 nm thick. The third coating was a nanocomposite sandwich coating formed by a TiSiN adhesive layer (50nm), a functional nanocomposite AlTiSiN layer (2 μm) and a top functional TiSiN layer with a thickness of 1 μm. The coatings were prepared by cathodic arc evaporation using the LARC® technology. After deposition, the samples were annealed at high temperatures (700 °C, 800 °C, 900 °C, 1000 °C) for one hour in air. With X-ray diffraction and scanning electron microscopy, the morphology, structure and phase composition of coatings and oxide layers after annealing were evaluated.
After the heat-treatment at 800 °C, cross-sectional and surface morphology images revealed, that an oxide layer has grown on top of the all coatings. EDS analysis showed that the oxide layers on the multilayers coatings had a layered structure, which was found to be the consequence of the high diffusion rate of Al. The interface with the coatings was very rough. It was depleted of Al, which diffused to the upper surface of the oxide. The oxidation process of nanocomposite AlTiSiN/TiSiN coating started by forming an oxide layer of ternary Ti–Si–O at the interface. This system segregated into two phases of TiO_2 and SiO_2. The nanocomposite coating remained stable even after annealing at 1000 °C. The surface of the multilayers was accompanied by a network of cracks at this temperature. An interesting phenomenon was also observed on the substrates itself during the annealing. As a result of oxidation, unprotected sides of the substrates expanded. Globular oxides based on tungsten and cobalt were also formed on the surface of the samples at the places where the coating was peeled off. The reasons for the differences are discussed.

1. Introduction

The AlTiN system has found a lot of industrial applications as hard and wear resistant coatings for cutting tools to enhance their life- time. These coatings with cubic NaCl (c) structure, where Al substitutes for Ti in the TiN based structure, are one of the most preferred wear resistant materials for advanced machining processes. Coatings combine very good mechanical and chemical properties, such as high oxidation resistance and high thermal and chemical stability. Thermal stability is a general term used to describe the changes occurring in the material as a function of temperature. These include oxidation, microstructure, composition or mechanical properties. This is a very important issue for industrial applications, as during the high-performance metal cutting operations, degradation of the coating occurs by wear related to oxidation or diffusion. Aluminum, present in the AlTiN system is an efficient additive improving oxidation and wear resistance and enhancing the thermal stability. Coatings can be applied up to temperatures of 800 °C. On the coating surface is grown a stable protective Al_2O_3 film in oxidizing environment. It is a highly insulating oxide, with low ion mobility and creates an effective oxygen diffusion barrier. Similar oxidation resistance was obtained by the addition of Si to TiN system where the protective barrier from oxygen diffusion is formed by amorphous oxides of silicon (SiO_2). TiSiN and AlTiSiN are nanocomposite coatings due to formation of amorphous phase of Si_3N_4 which is form around the AlTiN (TiN) crystals and also create barrier for oxygen diffusion [1-11].

2. Experimental details

2.1. Coating deposition

In this work the structure and morphological changes of the PVD coatings after isothermal annealing were investigated. The coatings were deposited on WC-Co (6 wt. % Co) cemented carbides discs with a 12 mm in diameter and a mirror-polished surface. The substrate samples were then ultrasonically cleaned for 15 minutes in acetone.

Three different coatings based on the AlTiN system were prepared by cathodic arc evaporation using the LARC® technology PLATIT π^{80}+DLC. Three combination of cylindrical rotating cathodes (Al, Ti, TiSi) are shown in Figure 1. Two coatings were multilayers with coarse-grained columnar structure with a chemically graded nanomultilayering. Both multilayers have an initial thickness of 3 µm. AlTiN/TiAlN consisting 20 layers – 10 layers of Al-rich AlTiN with 200 nm thick and 10 layers of Ti-rich TiAlN with 100 nm thick. For AlTiN/TiN multilayer was applied 10 AlTiN layers with the same thickness of 200nm and 10 layers of TiN with the 100 nm thick. The third coating was a nanocomposite sandwich coating formed by a TiSiN adhesive layer (50 nm), a functional nanocomposite AlTiSiN layer (2 µm) and a top functional TiSiN layer with a thickness of 1 µm. The current values and the chemical composition of the coating are shown in Table 1. The reactive gas was N and the chamber pressure was 1.9 Pa. The temperature during the deposition was 470 °C and the negative bias voltage on the substrates was −60V.

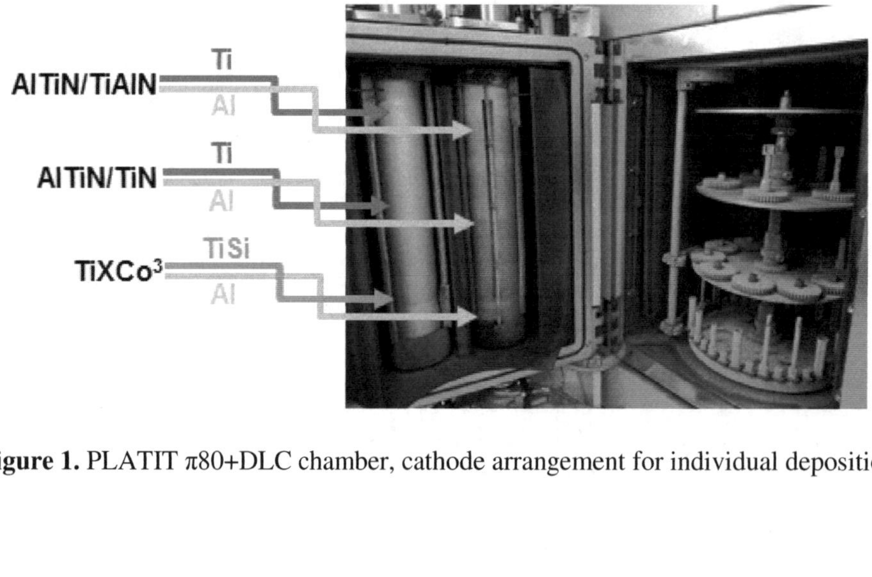

Figure 1. PLATIT π80+DLC chamber, cathode arrangement for individual deposition.

Table 1. Deposition DC current and chemical composition.

Coating		DC current [A]			Chemical composition [at. %]			
		Al	Ti	TiSi	N	Al	Ti	Si
AlTiN/TiAlN	AlTiN	162	95	-	55	27.5	17.5	-
	TiAlN	95	95	-	55	12.6	32.4	-
AlTiN/TiN	AlTiN	162	95	-	55	27.5	17.5	-
	TiN	-	150	-	53	-	47	-
TiXCo[3]	AlTiSiN	140	-	95	54.5	25.4	16.3	3.8
	TiSiN	-	-	150	57.2	-	33.2	8.6

2.2. Isothermal annealing

The oxidation resistance of the coatings was determined by an annealing process at temperatures of 700, 800, 900 and 1000 °C without a protective atmosphere in a laboratory annealing furnace for 1 hour, followed by continuous cooling in air. At 1000 °C, the holding time was only 20 minutes to avoid deformation of the substrate.

2.3. Chemical and structural characterization

The morphology and structure of the fracture areas of the individual coatings before and after their heating were observed by a JEOL JSM 7600F scanning electron microscope. Imaging was done using a secondary electron detector (LEI) or a backscattered electron detector (RBEI). Parameters during observation were: U = 20 keV, I = 2 nA and WD = 15 mm. Quantitative point and area analysis using a Si (Li) EDX detector X-Max 50 mm^2 from Oxford Instruments was used to identify the chemical composition of the coatings as well as oxidation products on the surface of the coatings. Using the same detector, changes in element content after the annealing processes from the surface of the coatings to the base material were also monitored by qualitative linescan EDX analysis.

Identification of phases present in the investigated coatings was performed by X-ray diffraction analysis using an Empyrean diffractometer from PANalytical with an anode CoKα (λ = 0.178897 nm), with Bragg-Brentan geometry (Θ - Θ), X-ray mirror to achieve primary parallel beam and positionally sensitive PIXcel3 detector. Analysis of individual samples was performed using GIXRD (Grazing Incidence X-ray Diffraction) geometry at a constant angle of incidence of a parallel X-ray beam to the sample surface (ω - 0.5 °). In this way, controlled penetration of X-ray radiation into the surface layers of the materials is only achieved, and thus this method is suitable for identifying the phase composition of the coatings. The parameters for the Co anode used during the measurement were as follows: U = 40 kV, I = 40 mA, angular interval 2θ (35 ÷ 100 °).

3. Results

3.1. Morphology and structure of the fracture areas

Figure 2 shows the morphology of all investigate coatings after deposition. Macroparticles of various sizes have been observed on the coatings, the formation of which is related to the deposition process. Also imperfections in the form of peeled parts of coatings can be seen. At a detailed magnification (5000x), an interesting network structure can be seen, which is present in all observed coatings, and its formation is probably due to the presence of plasma in the coating process.

The fracture areas of the coatings (Figure 3) show the samples structure perpendicular to the coatings. The individual layers of the coatings can be recognized by the material contrast of the RBEI detector. In the case of multilayers, the thinner TiAlN and TiN layers appear lighter due to the higher titanium content, the atomic number of which is higher than that of aluminum. On the contrary, thicker layers of AlTiN with higher aluminum content appear darker. The coatings are characterized by a columnar structure.

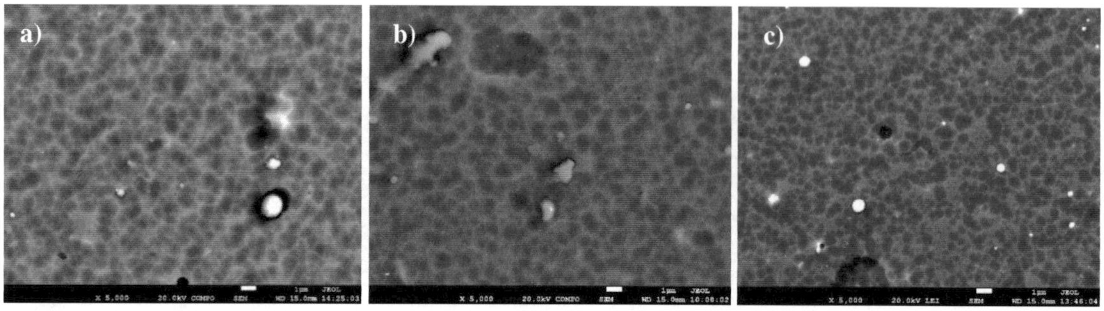

Figure 2. Morphology of the coatings a) AlTiN/TiAlN, b) AlTiN/TiN, c) TiXCo^{3}.

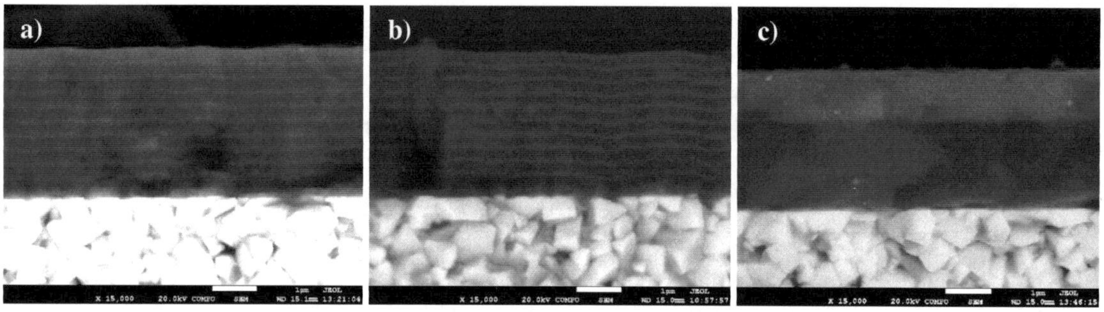

Figure 3. Fracture areas of the coatings a) AlTiN/TiAlN, b) AlTiN/TiN, c) TiXCo^{3}.

3.2. Isothermal annealing

3.2.1. AlTiN/TiAlN multilayer. The AlTiN/TiAlN coating was exposed to the air atmosphere at an elevated temperature of 700 °C for 1h. The coating morphology did not appear to have changed after heating. The occurrence of macroparticles as well as the network structure was maintained after annealing to the appropriate temperature. EDX qualitative point analysis was performed on the coating. The results show the presence of oxygen at 20 at.%, which indicates the formation of local oxides at a higher temperature but did not form a continuous oxide layer and the coating may be considered stable.

In Figure 4a, there is a macroscopic view of a sample that was annealed without a protective atmosphere at 800 °C. The coating changed color, indicating oxidation of the surface. It is also evident that after annealing, the substrate itself oxidized at the no-coated edges of the sample.

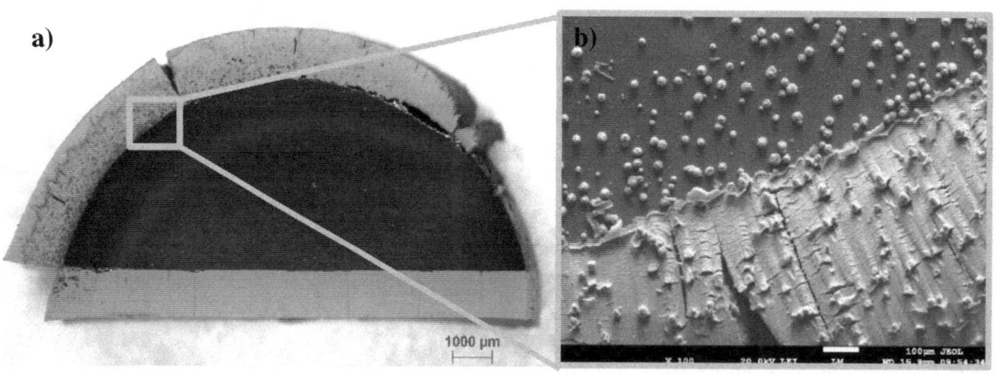

Figure 4. a) Oxidation of the WC-Co sample coated with AlTiN / TiAlN after annealing at 800 °C b) Edge of the coating with substrate oxides.

An oxide layer with a thickness of more than 1 mm was formed which lined the edges of the sample. By its expansion, the substrate damaged the coating on the edges, as can be seen in the detail of Figure 4b.

The morphology of the multilayer surface (Figure 5a) did not show signs of oxidation even after annealing to 800 °C. It can be stated that despite the expansion of the substrate at the edges of the sample, the layer retains integrity and protects the base material. At a larger magnification (Figure 5b) was not observe the classical fine network structure but the presence of oxides was noted. Their formation is the result of the early oxidation of the coating. The presence of oxygen in the coating has increased significantly by approximately 25 at. % compared to the results giving the chemical composition values after annealing at 700 °C.

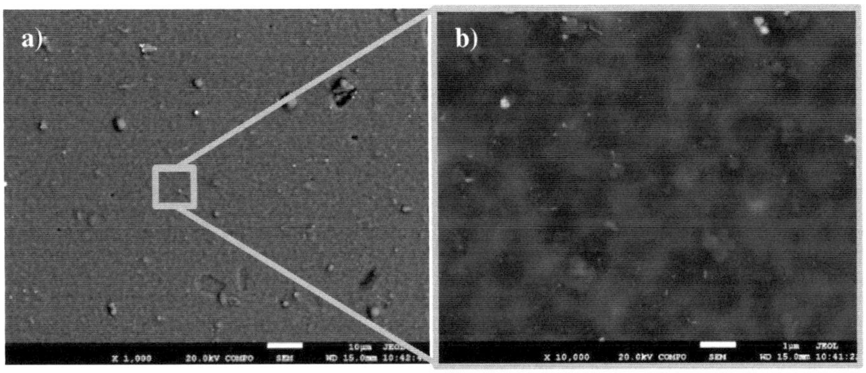

Figure 5. Morphology of the AlTiN/TiAlN multilayer after annealing at 800 °C a) 1000x, b) 10 000x.

Substrate oxidation was also confirmed by quantitative point EDX analysis. The measured values of the elements are given in Table 2, the points showing the area of analysis are documented in Figure 6. Spectrum 1 confirmed the presence of a damage coating at the edge of the substrate and spectrum 2 and 3 showed oxidation of the base material. Based on the literature and analyzed oxygen content (Table 2), it can be assumed that the oxidation products of the sintered carbide are probably oxides of WO_3 or MWO_4 (M = Co).

The top layer of the coating about 0.35 μm thickness has changed. It creates a different color contrast, which can confirm the theory that an oxide layer is beginning to form on the surface. However, the structure of the multilayer is still stable, free from cracks and with good adhesion to the base material. A linescan EDX analysis of the fracture area was also performed for analysis.

Figure 6. Substrate oxides after sample annealing at 800 °C.

Table 2. Results of point EDX analysis of substrate edge after annealing at 800 °C in at. %.

	C	O	Al	Ti	Co	W
Spektrum 1	5.16	70.37	5.77	11.88	1.66	5.16
Spektrum 2		73.78			8.23	17.99
Spektrum 3	21.62	53.02			11.79	13.57

In Figure 7 the course of the chemical composition from the surface of the coating to the substrate is shown. The overall graph shows that there is an increased concentration of oxygen on the surface of the coating, whose content decreases towards the coating. The thickness corresponds to the thickness of the top layer. The nitrogen concentration in this layer decreases. Also, an increased concentration of aluminum is recorded on the surface of the coating, suggesting that oxygen preferentially binds to aluminum resulting in the formation of aluminum oxides (probably Al_2O_3). The presence of this type of oxides on the surface of the coatings is considered to be a beneficial consequence of the oxidation process as they result in an increase in their wear resistance.

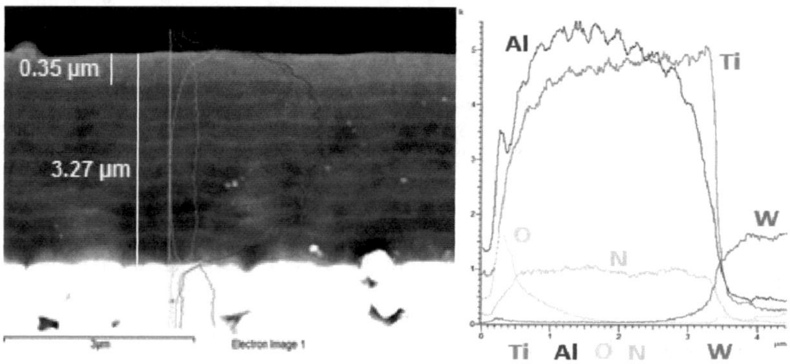

Figure 7. EDX linescan of fracture area of multilayer AlTiN/TiAlN after annealing at 800 °C.

After annealing the sample at 900 °C, it can be seen that the substrate, by its oxidation damage the multilayer at the edges of the sample as a previously case. The globular substrate oxides were also re-formed and became trapped at the edge of the coating (Figure 8a). At a small magnification, the multilayer morphology does not show significant signs of oxidation.

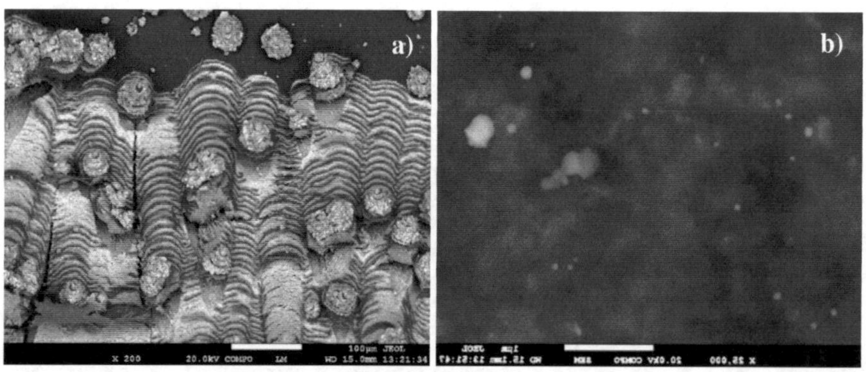

Figure 8. a) Edge of the coating with substrate oxides, b) AlTiN/TiAlN coating morphology after annealing at 900 °C.

On the surface there are sporadically visible substrate oxides, which are formed at delaminated areas of the coating where oxygen diffused into the exposed base material. At a larger magnification (Figure 8b), it is possible to see oxides on the surface of the multilayer.

EDX linescan analysis (Figure 9) confirmed the increased oxygen concentration at the surface of the coating and confirmed consisting of two layers. Aluminium oxides (Al_2O_3) are likely to be formed on the surface, and an oxide sublayer is formed beneath this layer, which appears to be a combination of titanium oxides (TiO_2) and aluminum oxides (Al_2O_3). The presence of nitrogen was observed again under the oxide layer, and its content was of heterogeneous character.

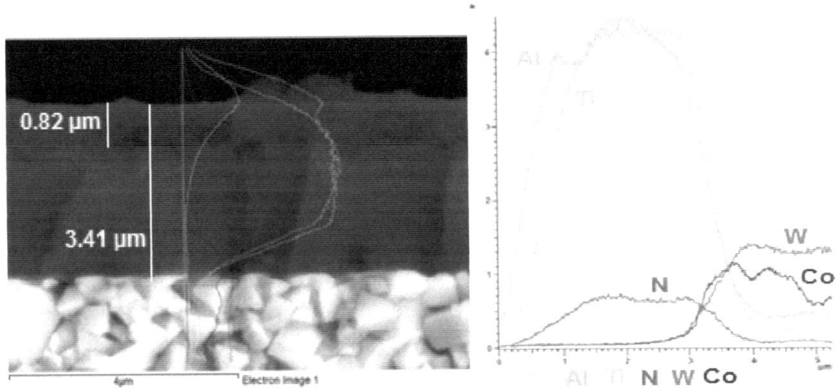

Figure 9. EDX linescan of fracture area of multilayer AlTiN/TiAlN after annealing at 900 °C.

After annealing at 1000 °C, a continuous oxide layer was formed on the AlTiN/TiAlN multilayer (Figure 10b). Cracks are visible on the entire surface of the coating (Figure 10a). The fracture area of the coating shows that the AlTiN/TiAlN coating is degraded to a considerable depth (Figure 11). The initial coating thickness of 3.2 µm was reduced to approximately 2.7 µm. A third, already from the reduced layer thickness, represents a continuous oxide layer.

The course of the EDX linescan analysis has not changed significantly from the previous case. Only the oxide layer/coating interface has advanced to a higher depth. Their course is shown in Figure 11. The oxide layer is predominantly composed of two types of oxides - Al_2O_3 and TiO_2.

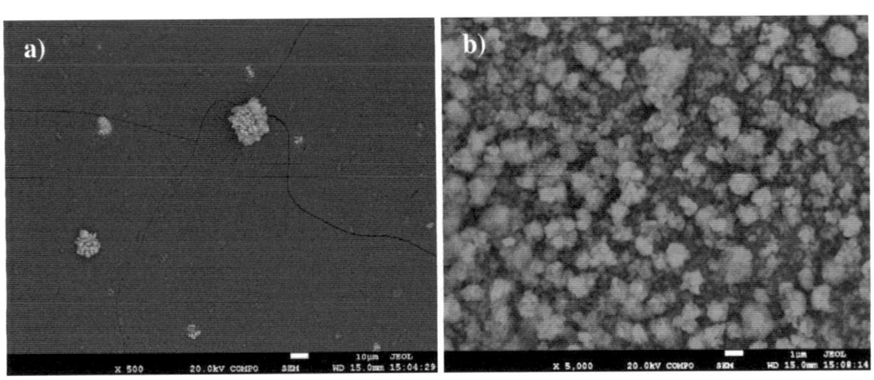

Figure 10. AlTiN/TiAlN coating morphology after annealing at 1000°C a) 500x, b) 5 000x.

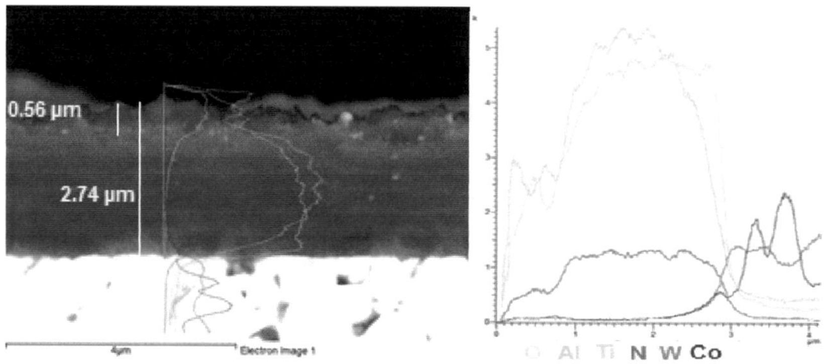

Figure 11. EDX linescan of fracture area of multilayer AlTiN/TiAlN after annealing at 1000 °C.

3.2.2. AlTiN/TiN multilayer

The development of the AlTiN/TiN coating after annealing at individual temperatures was very similar as in case of the previous multilayer. The morphology of the coating after heating to 700 °C is not different from the morphology before annealing. It contains the macroparticles that were formed during the deposition process and also a typical network structure. Substrate oxidation occurs at 800 °C. According to Figure 12b, it is evident that it is the same mechanism as the AlTiN/TiAlN coating. The globular substrate oxides increased by scrolling and captured on the coating edges, which were damaged by the expansion of the cemented carbide due to the elevated temperature.

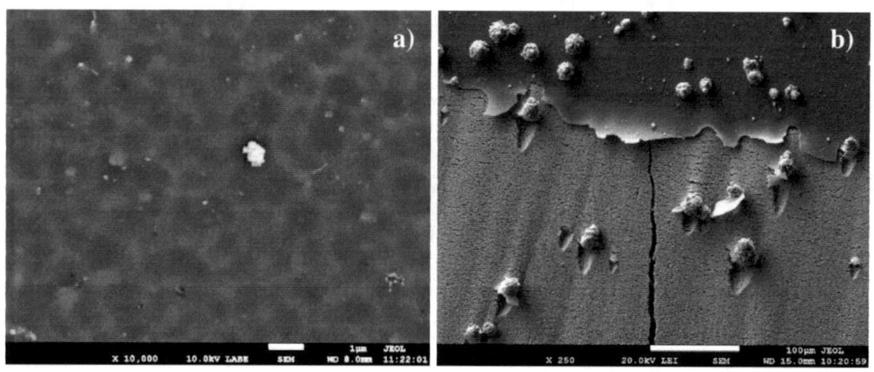

Figure 12. a) Morphology of the AlTiN/TiN multilayer after annealing at 800 °C, b) Edge of the coating with substrate oxides.

By means of EDX analysis, a qualitative linescan analysis was also performed on the fracture area. The analysis results are shown in Figure 13. The highest amount of oxygen is recorded at the top of the coating and decreases towards the substrate. The depth of its highest concentration corresponds to the thickness of the top layer. It can also be seen from the change in concentration that the top layer is depleted of titanium and nitrogen as a result of diffusion. Apart from oxygen, is predominantly aluminum on the top, and thus is believed to formation of an Al_2O_3 passivation layer. The interface of the coating-substrate system is intact, the layer is stable and the adhesion has not deteriorated.

The edges of the sample after heating to 900 °C have the same oxidation behavior as in the previous multilayer (Figure 14). By oxidizing the substrate tears the layers and damages the edges of the coating. The coating morphology is characterized by the presence of small crystals of oxides. The initial coating thickness - 3.1 µm was reduced by 0.4 µm. The missing portion of the coating was degraded by oxidation. The top of the coating (approximately 0.2 µm) is oxidized. However, the remainder of the coating up to the substrate is stable as in the previous cases. Oxygen did not diffuse to a depth of more than 0.8 µm.

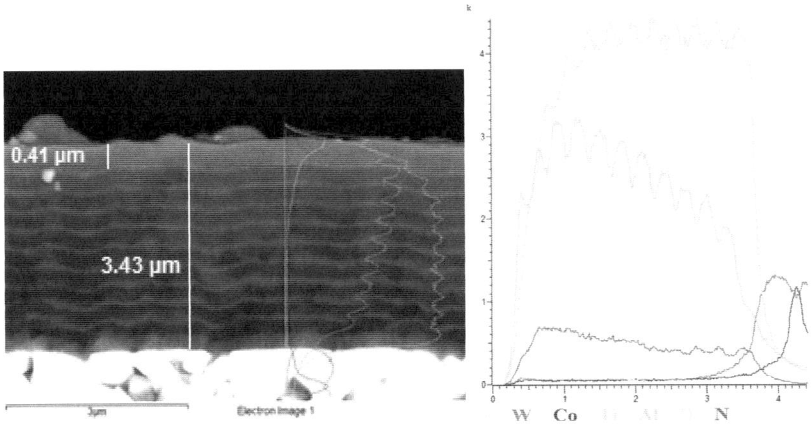

Figure 13. EDX linescan of fracture area of multilayer AlTiN/TiN after annealing at 800 °C.

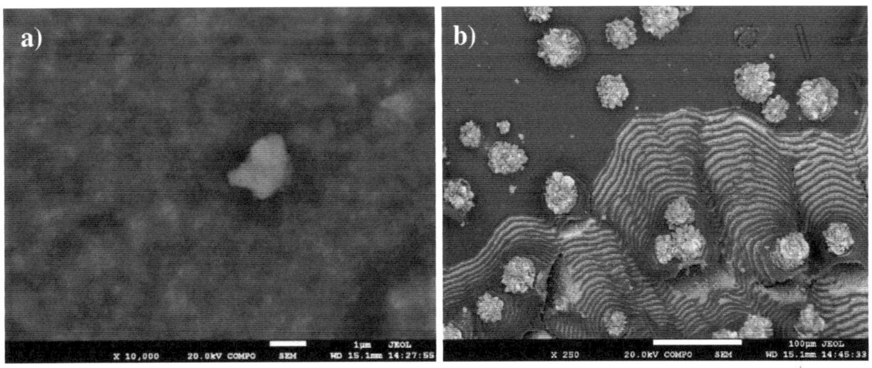

Figure 14. a) Morphology of the AlTiN/TiN multilayer after annealing at 900 °C, b) Edge of the coating with substrate oxides.

After annealing the coating at 1000 °C, the oxide layer of the cemented carbide reached a thickness of more than 1.5 mm. The multilayer analyzed after oxidation had the same behavior as the AlTiN/TiAlN multilayer. A large amount of oxides and cracks was present on the surface coating (Figure 15a).

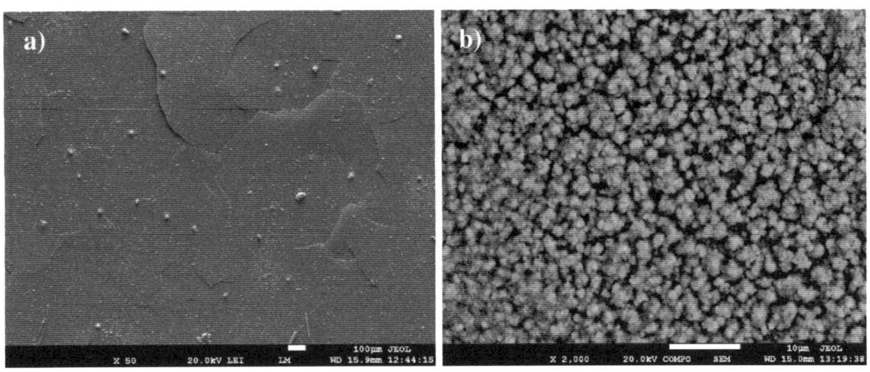

Figure 15. AlTiN/TiN coating morphology after annealing at 1000 °C a) 50x, b) 2000x.

From the fracture area of the multilayer, is obvious that its thickness has dropped by almost 0.8 µm and the oxidized layer of the remaining coating is also 0.8 µm but has no homogeneous thickness over the entire coating (Figure 16).

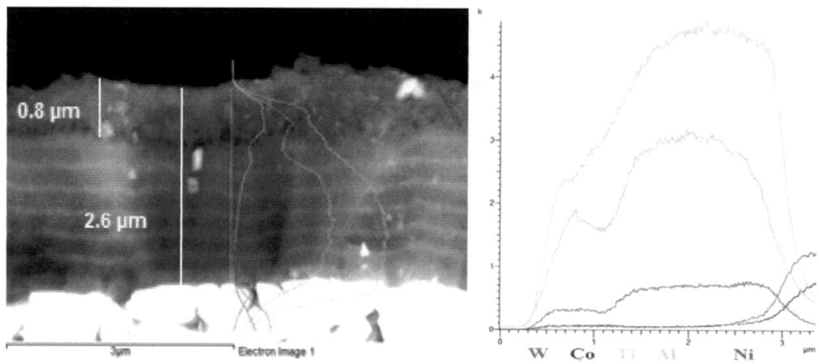

Figure 16. EDX linescan of fracture area of multilayer AlTiN/TiN after annealing at 1000 °C.

3.2.3. TiXCo³. The TiXCo³ nanocomposite coating retained its morphology as it had before heating to 700 °C. Using quantitative point EDX analysis, the surface showed the presence of oxygen (approximately 40 at.%). However, there is no continuous oxide layer on the coating; it is the formation of local oxides. Once the sample has been heated to 800 °C, the substrate is also oxidized. The oxide layer of cemented carbide has the same character as in the previous two cases of multilayers. The mechanism of formation of globular oxides of sintered carbide also has two methods of formation in this coating. The first is the scrolling of the oxides towards the coating and subsequent capture at its edge. The second mechanism is the formation of oxides at places where the coating had a damaged structure (peeled off parts of coatings due to the residual stresses) and oxygen could diffuse directly into the substrate. Figure 17 b documents the substrate oxide formed at the peeled part of the coating. Coating residues are visible on the oxide. The morphology of the coating after heating to 800 °C already shows signs of oxidation. In Figure 17a, small crystalline formations are seen, which are apparently titanium oxides (TiO_2). The fracture area showed that the sandwich nanocomposite coating increased its thickness by 0.4 µm after annealing. This is obviously an oxide layer containing titanium oxides, preferably TiO_2, but the formation of silicon oxides (SiO_2) is possible. Interestingly, unlike multilayers, the oxide layer has grown on the coating and has not oxidized the top coating layer.

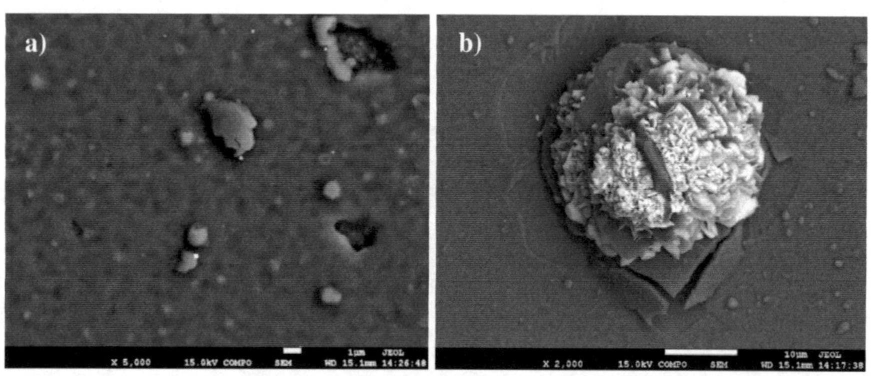

Figure 17. a) TiXCo³ coating morphology after annealing at 800 °C, b) Substrate oxide on the coating surface.

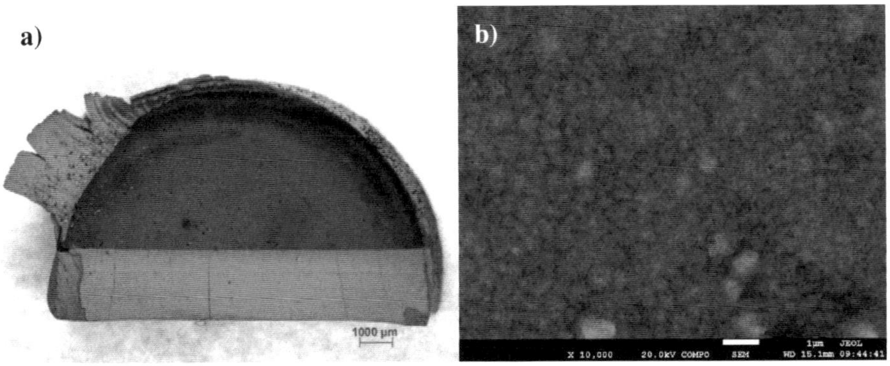

Figure 18. a) Oxidation of the WC-Co sample coated with TiXCo³ after annealing at 900 °C
b) TiXCo³ coating morphology after annealing at 900 °C.

A TiXCo³ coated sample after annealing at 900 °C is shown in Figure 18a. The substrate underwent oxidation and formed an oxide layer up to 2.5 mm thick. The oxide layer was formed in layers. The coating morphology was oxidized as in the previous case. A larger magnification indicates that the coating is coated with a continuous layer of oxides having a different morphology than that of multilayers (Figure 18b). On the fracture area of the sandwich nanocomposite coating, it is possible to observe a continuous oxide layer formed on its surface. While after annealing at 800 °C, it appeared that the oxide layer was formed on the coating and did not affect TiXCo³ alone, in Figure 19 it can be seen that the thickness of the top layer decreased in this case and oxidation also occurred in the coating itself. The total layer thickness after heating at 900 °C is 3.59 µm and the oxide layer has reached 0.74 µm.

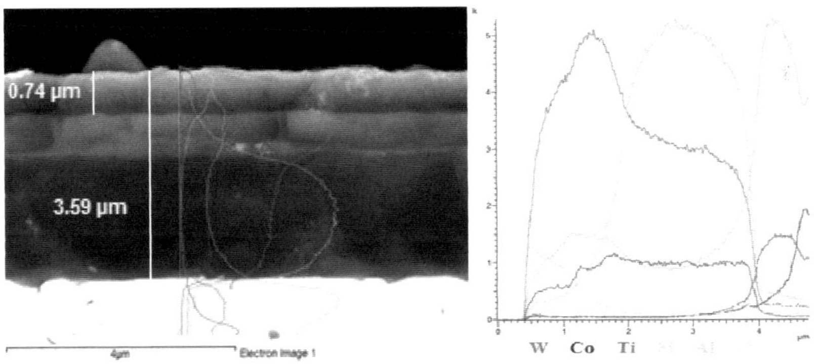

Figure 19. EDX linescan of fracture area of TiXCo³ after annealing at 900 °C.

After annealing at 1000°C there is a continuous oxide layer on the TiXCo³ coating (Figure 20a). This is confirmed by a quantitative EDX analysis which recorded an amount of oxygen of about 70 at.% on the surface. Occasionally substrate oxides are also present on the oxide layer. In Figure 20b, the substrate oxide formed in damaged areas of the coating is shown. The globular oxide gradually disrupted the coating during its growth, causing an interesting morphology around the oxide.

The fracture area of the coating (Figure 21) has an interesting structure. The total coating thickness, due to the oxide layer, increased by 1.38 µm. The nanocomposite functional layer did not change its thickness, but the top layer thinned by 0.5 µm. The oxide layer consists of two different layers. The first one is 0.63 µm thick and the second one is 0.74 µm thick. By EDX linescan analysis it is not possible to determine exactly what type of oxides it is. Both layers have approximately the same chemical composition, but the lighter layer has a higher proportion of O and less Ti and Si.

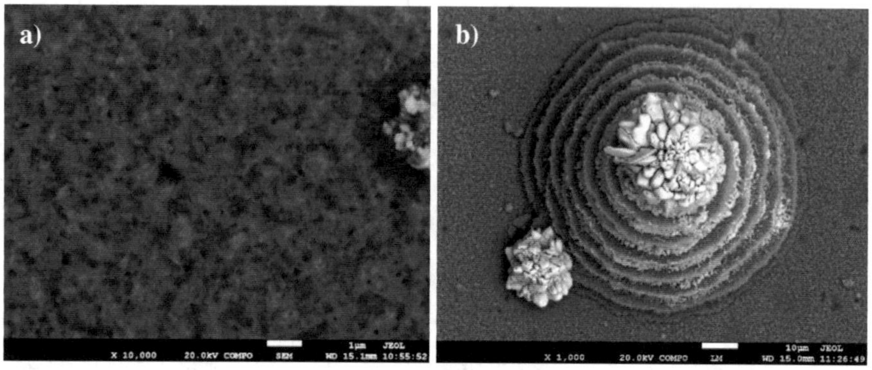

Figure 20. a) $TiXCo^3$ coating morphology after annealing at 1000 °C, b) Substrate oxide on the coating surface.

However, the coating protected the sintered carbide even when heated to the highest temperature of 1000 °C. It should also be pointed out that the functional and adhesive layer resisted by oxidation and remained intact. At 1000 °C, the $TiXCo^3$ nanocomposite coating is able to protect its substrate from oxidation.

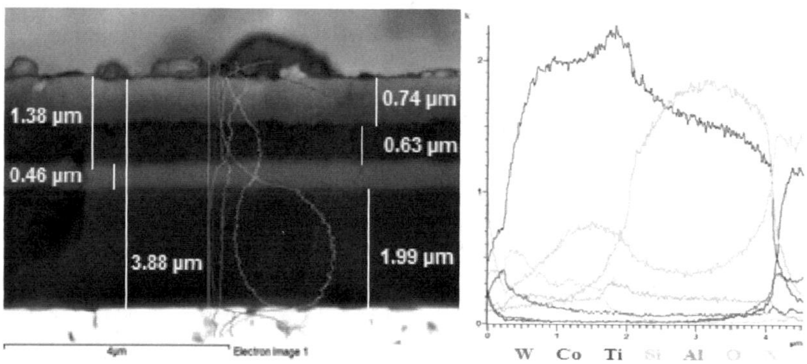

Figure 21. EDX linescan of fracture area of $TiXCo^3$ after annealing at 1000 °C.

3.3. Phase composition analysis

Qualitative X-ray diffraction analysis confirmed the presence of one phase in all three deposited layers. It is the cubic phase of TiN, whose preferred orientation was in the crystallographic plane (002). In addition to this plane, the TiN phase also diffracted significantly on planes (111) and (022), also for all three coatings. For both multilayers, a shift to the right of the diffraction maxima for planes (002) and (022) was noted, which represents a reduction in the lattice parameter due to the substitution of aluminum atoms into the lattice of the cubic solid TiN solution. In addition to the phase present in the coatings, two phases belonging to the base material were identified. It is the hexagonal phase of WC diffracting on planes (010), (011) and (001) and hexagonal phase Co.

X-ray diffraction analysis was also used to identify oxidation products, by analyzing selected samples at various heating (AlTiN/TiAlN multilayer after heating to 800 °C, multilayer AlTiN/TiN after heating to 900 °C, multilayer AlTiN/TiAlN after heating to 1000 °C and nanocomposite $TiXCo^3$ sandwich coating after heating to 1000 °C). Given the thickness of the coatings and especially the thickness of the oxide layers, this analysis was also carried out in GIXRD geometry. From the diffraction pattern shown in Figure 22a was observed that the oxidation of the surface layers was already at 800 °C.

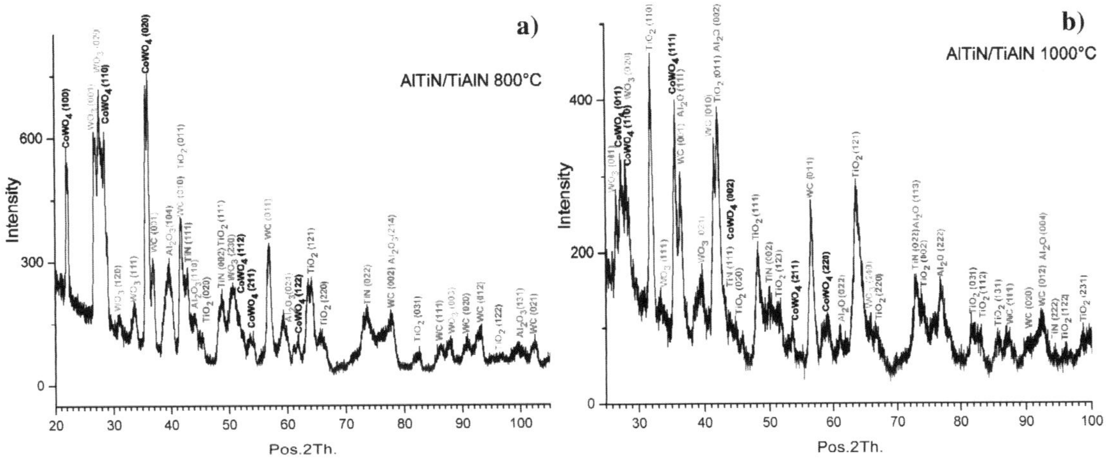

Figure 22. Diffraction pattern of a) AlTiN/TiAlN coating annealed at 800 °C and b) at 1000 °C after GIXRD analysis.

The hexagonal Al_2O_3 phase and the tetragonal TiO_2 phase were identified as oxidation products and the cubic Al_2O phase was detected in the case of the AlTiN/TiAlN multilayer annealed at 1000 °C. However, in the case of oxides, diffraction occurs over a wide-angle range, with the individual angles being very similar, and therefore it is not excluded that other types of Al and Ti-based oxides may have formed on the surface of the coatings. In addition to the layer oxides, the oxides of the base material in the form of oxides WO_3 and $CoWO_4$ from the edges of the samples were also detected (Figures 22, 23).

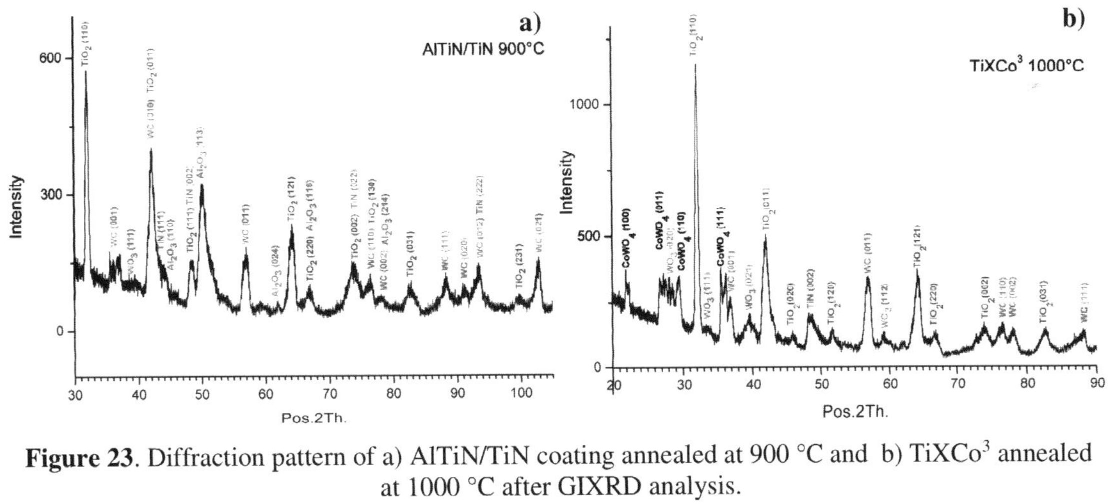

Figure 23. Diffraction pattern of a) AlTiN/TiN coating annealed at 900 °C and b) TiXCo^3 annealed at 1000 °C after GIXRD analysis.

4. Discussion and conclusion

As a result of isothermal annealing, continuous oxide layers were formed already at 800 °C. For both multilayers, progressive oxidation evolution was the same by increasing temperatures. The formation of an oxide layer was predominantly made up of Al_2O_3 and TiO_2 oxides. Wu and Lee, who investigated the oxidation resistance of AlTiN-based coatings, confirmed the presence of these oxides. [12-13] After performing X-ray diffraction analysis showed the presence of hexagonal phase Al_2O_3 and tetragonal phase TiO_2, but the presence of other types of oxides based on Ti and Al was not excluded. Titanium Dioxide - Rutile is a nanostructured phase with the space group P 42/m n m and crystallographic parameters $a = b \neq c$, $\alpha = \beta = \gamma$ (90°, 90°, 90°). Aluminium Oxide (Al_2O_3) is a hexagonal phase with the space group R -3 c and crystallographic parameters $a = b \neq c$, $\alpha = \beta \neq \gamma$ (90°, 90°, 120°). Al_2O_3 passivation layer was formed on the surface. This phenomenon occurs in many literature publications that focus on AlTiN-based coatings analyzes. This formation was also confirmed in this case. Due to its slow ion mobility, the layer acts as an effective barrier to oxygen diffusion. In addition to acting as a protective coating, it also improves mechanical properties due to its hardness. However, based on analyzes, it can be concluded that the layer is only stable up to a certain temperature. After annealing at 900 °C, the multilayers gradually degraded. Their thickness decreased and the thickness of the oxide layer increased. Nevertheless, the coating showed stability.

When the coating was heated to 1000 °C, the degradation of the multilayers was complete. In addition to the fact that the coatings have lost a considerable part of their thickness as a result of oxidation, the surface has cracked and thus completely degraded.

The sandwich nanocomposite coating had a different course of oxide layer formation. At 700 °C the oxygen content of the coating was increased and at higher temperatures small crystals of oxides were already formed. However, these oxides were not formed at the expense of the coating but as an additional layer on the coating. At higher temperatures, the continuous oxide layer has already penetrated into the coating. The different course of oxidation is probably due to the presence of Si. In his study, Choi examined the effect of silicon on the formation of an oxide layer and found that the higher the silicon content of the coating, the higher the oxidation resistance. According to studies by Choi and Zhang, the presence of TiO_2 or SiO_2 oxides was suggested [14-15]. X-ray diffraction analysis confirmed only the presence of titanium oxides, since SiO_2 is characterized by an amorphous structure and its formation on the coating is therefore not excluded. Unlike multilayers, the nanocomposite coating has maintained its stability even at 1000 °C. No degradation due to cracking. The coating thickness did not decrease significantly. An interesting feature was the formation of a two-layered oxidic film that was formed at this temperature. Segregation of titanium and silicon oxides is believed to have occurred. Silicon oxides act as a high temperature barrier to the diffusion of oxygen.

In addition to the oxidation of the coatings themselves, the WC-Co base material was also oxidized. The edges of the samples were not protected by coatings and therefore oxygen was directly exposed to the substrate, causing its oxidation. As a result, the edges of the samples began to expand and the coating on the edges was damaged. Globular oxides have also been observed which have been attached to the edges of the coatings by scrolling. Their occurrence has also been documented on surface of coatings. Their formation is associated with coating defects. Due to the high stress in the coatings, small parts of the coating are peeled off during the deposition process. In some cases, the delamination will completely uncover the base material. Such sites are then exposed to an air atmosphere at elevated temperatures and the substrate is oxidized. Substrate oxides increased at these sites and were observed in the form of globular formations. X-ray diffraction analysis showed that they are oxides of WO_3 and $CoWO_4$, which they also reported in their studies by Ghadami and Geng. [16-17] Although the coating protects the substrate at elevated temperatures, the presence of these local oxides in the weakened areas is not desirable as they may tear off and damage the layer in the machining process.

Acknowledgments
This publication was supported by the Project of VEGA Grant Agency of the Ministry of Education, Science, Research and Sport of the Slovak Republic and Slovak Academy of Sciences, No. 1/0540/19: “Research of possibilities of increasing the thermal and oxidation stability of Al-Ti-N based hard coatings” and also by the Slovak Research and Development Agency with grant number APVV-15-0168: “Research of modification of phase interfaces in the coating/substrate system to increase the adhesion of hard coatings” and the Operational Program Research and Innovation for the project: "Independent research and development of new superhard coatings and their characterization by advanced experimental techniques" (ITMS code: NFP313010T598), co-financed by the European Regional Development Fund.

References

[1] Parlinska-Wojtan M 2016 Oxygen diffusion in columnar TiAlSiN coatings investigated by electron microscopy *Thin Solid Films* **616** 437-443

[2] Chang Y Y and Yang S M 2010 High temperature oxidation behavior of multicomponent TiAlSiN coatings *Thin Solid Films* **518** (21) S34-S37

[3] Chen L, Xu Y X, Du Y and Liu Y 2015 Effect of bilayer period on structure, mechanical and thermal properties of TiAlN/AlTiN multilayer coatings *Thin Solid Films* **592** Part A 207-214

[4] PalDey S and Deevi S C 2003 Single layer and multilayer wear resistant coatings of (Ti,Al)N *Mater. Sci. Eng.* **A 342** 58-79

[5] Mclntyre D, Greene J E, Hakansson G, Sundgren J -E and Münz W –D 1990 Oxidation of metastable single-phase polycrystalline $Ti_{0.5}Al_{0.5}N$ films: kinetics and mechanisms *J. Appl. Phys.* **67** 1542–1553

[6] Hörling A, Hultman L, Odén M, Sjölen J and Karlsson L 2005 Mechanical properties and machining performance of $Ti_{1-x}Al_xN$-coated cutting tools *Surf. Coat. Tech.* **191** (2-3) 384-392

[7] Mayrhofer P H, Hörling A, Karlsson L, Sjölen J, Larsson T, Mitterer C and Hultman L 2003 Self-organized nanostructures in the Ti–Al–N system *Appl. Phys.* **83** 2049-2051

[8] Parlinska-Wojtan M, Karimi A, Coddet O, Cselle T and Morstein M 2004 Characterization of thermally treated TiAlSiN coatings by TEM and nanoindentation *Surf. Coat. Tech.* **188–189** 344

[9] Hofmann S 1990 Formation and diffusion properties of oxide films on metals and on nitride coatings studied with Auger electron spectroscopy and X-ray photoelectron spectroscopy *Thin Solid Films* **193/194** 648

[10] Jarms C, Stock H –R and Mayr P 1998 Mechanical properties, structure and oxidation behaviour of $Ti_{1-x}Al_xN$-hard coatings deposited by pulsed dc plasma-assisted chemical vapour deposition (PACVD) *Surf. Coat. Tech.* **108–109** 206-210

[11] Knotek O, Munz W D and Leyendecker T 1987 Industrial deposition of binary, ternary, and quaternary nitrides of titanium, zirconium, and aluminum *J. Vac. Sci. Technol. A* **5** 2173–2179

[12] Wu Z T, Sun P, Qi Z B, Wei B B and Wang Z C 2017 High temperature oxidation behavior and wear resistance of $Ti_{0.53}Al_{0.47}N$ coatings by cathodic arc evaporation *Vacuum* **135** 34-43

[13] Lee K S and Seo S M 2005 Oxidation behaviors of TiAl(La)N coatings deposited by ion plating *Scripta Materialia* **52** 445-448

[14] Choi J B, Kurn CH, Lee M H and Kim K H 2004 Effect of Si content and free Si on oxidation behavior of Ti-Si-N coatings layers *Thin Solid Films* **447** 365-370

[15] Zhang C H, Lu X C , Wang H, Luo J B, Shen Y G and Li K Y 2006 Microstructure, mechanical properties, and oxidation resistence of nanocomposite Ti-Si-N coatings *App. Surf. Sci.* **252** 6141-6153

[16] Ghadami F, Ghadami S and Abdollah-pour H 2013 Structural and oxidation behavior of atmospheric heat treated plasma sprayed WC-Co coatings *Vacuum* **94** 64-68

[17] Geng Z, Hou A, Shi G, Duan S and Li A 2016 Tribological behavior at various temperatures of WC-Co coatings prepared using different thermal spraying techniques *Tribolog Int.* **104** 36-44

IOP Conf. Series: Materials Science and Engineering **726** (2020) 012002 doi:10.1088/1757-899X/726/1/012002

Analysis of temperature and stress-strain fields during laser beam welding of a TRIP steel

Š Vrtiel and M Behúlová

Slovak University of Technology in Bratislava, Faculty of Materials Science and Technology in Trnava, Ulica Jána Bottu 2781/25, 917 24 Trnava, Slovak Republic

stefan.vrtiel@stuba.sk, maria.behulova@stuba.sk

Abstract. Transformation Induced Plasticity (TRIP) steels belong to the group of Advanced High Strength Steels (AHSS) that are characterized by good strength-strain combination and formability required for special applications in the automotive industry. The excellent combination of strength, ductility and formability of TRIP steels is achieved by a careful control of microstructure development in the process of their production. The microstructure of multiphase TRIP steels typically consists of ferrite, carbide-free bainite and metastable retained austenite, which can be transformed into martensite by plastic deformation. Upon crash, this feature allows the TRIP steels to absorb more energy, ensuring greater passenger safety. In the automotive industry, TRIP steels are mainly joined by resistance spot welding, laser or electron beam welding. Generally, the thermal cycle of a fusion welding process destroys the sophistically designed microstructure of these steels in fusion and heat-affected zones resulting in deterioration of mechanical properties of the weld joint. Negative consequences of the welding process can be eliminated using proper welding parameters. The paper deals with numerical simulation and analysis of the temperature and stress-strain fields developed during the laser beam welding of a CMnSiNb TRIP steel sheets with the thickness of 2 mm. Simulation model takes into account non-linear temperature and phase dependent material properties. The heat input during the laser beam welding is modelled using the conical volumetric heat source. The optimal welding parameters for production of butt joints of CMnSiNb TRIP steel sheets using the TruDisk 4002 disc laser with the maximum power of 2 kW are designed.

1. Introduction

Effort to reduce the fuel consumption and the weight of passenger cars leads to increased applications of advanced high-strength steels (AHSS). The use of AHSS in the automotive industry increases also the impact resistance of cars, ensuring greater passenger safety. However, maintaining a high degree of ductility of materials for the forming processes of car body components along with the effort to enhance the tensile strength requires the development of new materials [1-3]. These requirements led to the development of novel multiphase steels, including TRIP steels [4-5]. In the framework of the programme ULSAB - the Ultra Light Steel Auto Body, sheet steel producers from 35 companies and 18 countries cooperated on the research and production of new grades of AHSS, resulting in 25% decrease of weight and 14 % decrease of cost for the benchmarked four-door sedans [6].

The problem that arises in the area of application of AHSS and especially TRIP steels in the automotive industry is the formation of high-quality joints between the car body parts [7-8]. Welding represents currently the most applied technology for joining car body components.

TRIP steels are mainly joined by resistance spot welding, laser or electron beam welding [9-12]. Generally, the thermal cycle of a fusion welding process destroys the sophistically designed microstructure of these steels in fusion and heat-affected zones, resulting in deterioration of mechanical properties of the weldments. In the process of laser beam welding (LBW), the temperature cycles are very short, so the material is exposed to maximum temperatures for a much shorter time compared to e. g. arc or resistance spot welding. Moreover, laser welding is also characterized by the formation of a narrow heat affected zone (HAZ). These advantages make LBW technology useful for welding difficult to weld materials such as TRIP steels. It has been proven that the setting of optimal technological parameters can ensure production of satisfactory weld joints of TRIP steels or even combinations of TRIP steels with another materials [8-16].

The paper is focused on the numerical simulation and analysis of the temperature and stress-strain fields developed during the laser beam welding of CMnSiNb TRIP steel sheets with the thickness of 2 mm. Using developed simulation model, appropriate welding parameters for production of butt joints of CMnSiNb TRIP steel sheets using the TruDisk 4002 disc laser with the maximum power of 2 kW are designed.

2. Problem description and simulation model

Analysis of temperature and stress-strain fields during laser welding of sheets made of the CMnSiNb TRIP steel with the chemical composition given in Table 1 was performed using the program code ANSYS Release 18.1 [17]. The application of continuous wave TRUMPF TruDisk 4002 disc laser with the maximum power of 2 kW, the wavelength of laser radiation of 1030 nm and the laser core diameter of 400 μm was supposed to produce butt joints of steel sheets with the thickness of 2 mm.

In the first step, simulation model for the thermal and static analyses of the laser beam welding process was developed.

Table 1. Chemical composition of the TRIP steel.

Element	C	Mn	Si	P	S	Cr	Ni	Cu	Al	Nb	Mo	Fe
[wt. %]	0.21	1.449	1.797	0.008	0.005	0.008	0.072	0.058	0.006	0.06	0.02	bal.

2.1. Mathematical model

Numerical simulation of temperature fields is based on the solution of Fourier-Kirchhoff partial differential equation in the form [18]

$$\rho c_{\mathrm{p}} \frac{\partial T}{\partial t} = \nabla.(\boldsymbol{\lambda} \nabla T) + q_{\mathrm{v}}. \tag{1}$$

in which T is the temperature, ρ is the density, c_{p} is the specific heat, $\boldsymbol{\lambda}$ is the tensor of thermal conductivities and q_{v} is the volumetric heat source density, i. e. the heat generated per unit time in a unit volume. Geometrical, thermo-physical, initial and boundary conditions are necessary to define to accomplish solution of the Eq. 1 and numerical simulation.

Analysis of stress-strain fields during laser beam welding is based on the equilibrium condition given by the equation [19-20]

$$\nabla.\overline{\sigma} + \boldsymbol{f} = \boldsymbol{0} \tag{2}$$

where $\overline{\sigma}$ is the stress tensor and $\boldsymbol{f}$ is the vector of external volumetric forces. The constitutive model describing the mechanical behavior of TRIP steels generally supposes that the total strain rate $\dot{\varepsilon}^{tot}$ is given as the sum of the elastic $\dot{\varepsilon}^{e}$, plastic $\dot{\varepsilon}^{p}$ and transformation $\dot{\varepsilon}^{tr}$ parts

$$\dot{\varepsilon}^{tot} = \dot{\varepsilon}^{e} + \dot{\varepsilon}^{p} + \dot{\varepsilon}^{tr}. \tag{3}$$

Stress and strain tensors are interrelated through material properties: Young modulus and Poisson ratio. The thermal stresses are dependent on the coefficient of thermal expansion. A detail description of the constitutive model can be found in [19-21].

IOP Conf. Series: Materials Science and Engineering **726** (2020) 012002 doi:10.1088/1757-899X/726/1/012002

2.2. Geometrical and finite element model

Taking into account geometrical symmetry and symmetry of boundary conditions, the proposed geometrical model consists of a sheet with the dimensions of 15 mm × 30 mm × 2 mm (Figure 1). The finite element mesh was generated in the program code ANSYS, Release 18.1 [17] using 3D elements of the SOLID70 type (thermal analysis). For structural analysis, the element type was switched to SOLID185. The progressive mesh density was defined with respect to the expected different temperature gradients directly in and near the area of the weld joint and in the distant regions of the thermally unaffected base material. The highest mesh density was near the weld centreline where the element size was 0.1 mm in the welding direction, 0.05 mm in the direction perpendicular to the welding line and 0.125 mm along the sheet thickness. The generated finite element model consists of 110019 nodes and 140968 elements.

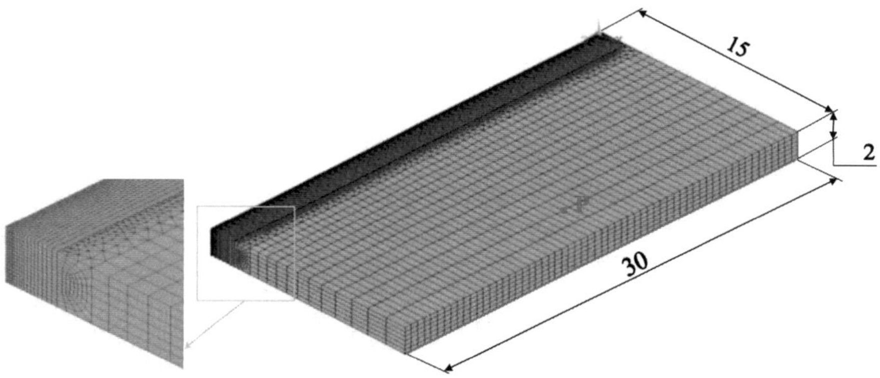

Figure 1. Geometrical model and finite element model with a detail of the mesh density near the weld centreline.

2.3. Material model

Microstructure and material properties of the applied CMnSiNb TRIP steel are influenced by the processing conditions of its thermo-mechanical treatment [22]. Mechanical and thermo-physical properties of the CMnSiNb TRIP steel in the dependence on temperature were computed using the JMatPro software [23]. Obtained thermal properties - values of the thermal conductivity, density and specific heat *v. s.* temperature are plotted in Figure 2. The solidus and liquidus temperatures of the CMnSiNb TRIP steel were computed to be 1422.4 °C and 1481.7 °C, respectively.

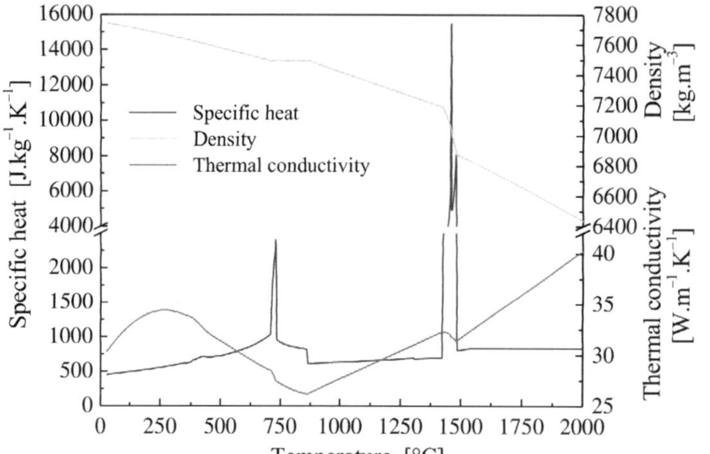

Figure 2. Thermal properties of the CMnSiNb TRIP steel in the dependence on temperature.

The latent heat associated with the solid-liquid phase transformation (l_v = 297.8 kJ/kg) was taken into account using the method of equivalent specific heat capacity [17, 24].

Mechanical properties including the Young modulus, Poisson ratio and the thermal expansion coefficient in the dependence on temperature are depicted in Figure 3. Non-linear stress-strain curves for chosen temperatures applied for static analysis are shown in Figure 4.

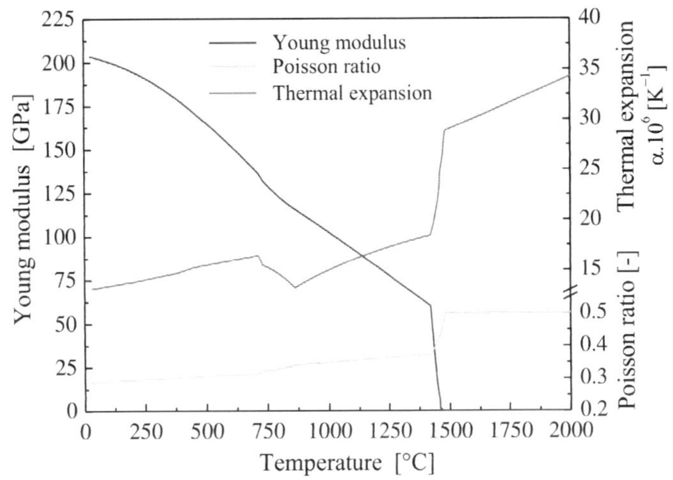

Figure 3. Mechanical properties of the CMnSiNb TRIP steel in the dependence on temperature.

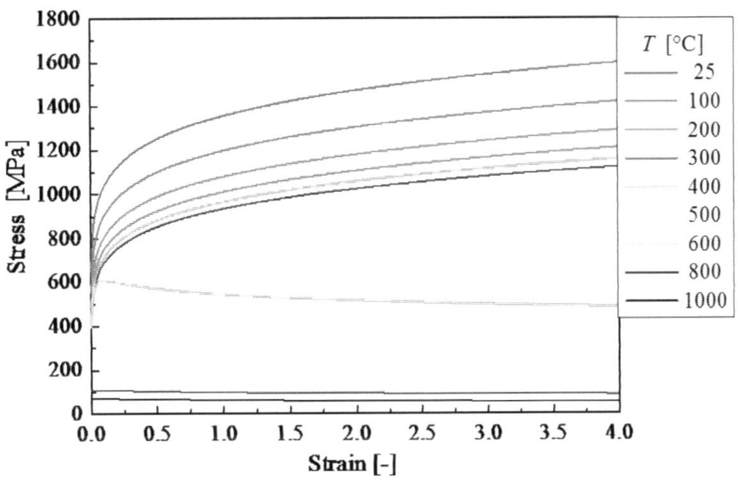

Figure 4. Stress-strain curves of the CMnSiNb TRIP for chosen temperatures.

2.4. *Initial conditions, boundary conditions and loads*

The initial temperature of the welded sheets was 20 °C. In performed numerical simulations, the heat extraction due to the convection and radiation to the surroundings was taken into account [25]. To model the heat input due to the moving laser beam source, the conical volumetric model of a heat source with the Gaussian heat flow density distribution was exploited (Figure 5).

$$q_v(x,y,z) = \frac{9\eta P e^3}{\pi(e^3-1)} \times \frac{1}{(y_e-y_i)(r_e^2+r_e r_i+r_i^2)} \exp\frac{3(x^2+z^2)}{r_0^2(y)} \tag{4}$$

$$\text{where} \quad r_0(y) = r_e + \frac{r_i - r_e}{y_i - y_e}(y - y_e) \tag{5}$$

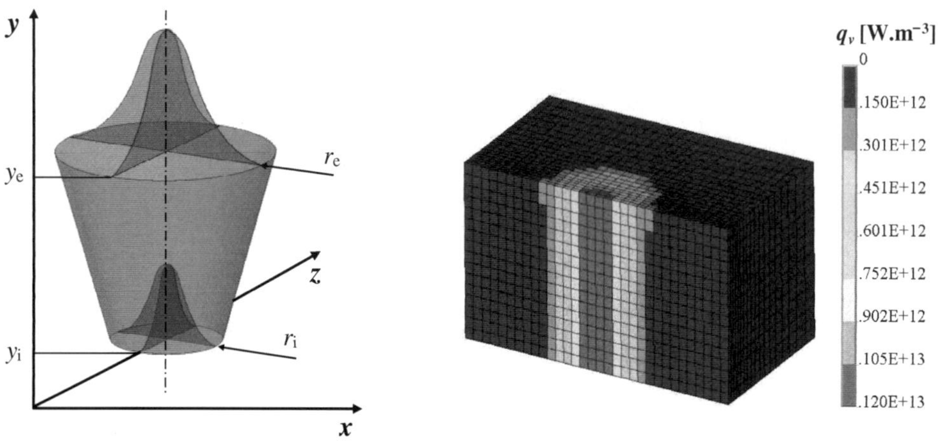

Figure 5. Conical volumetric heat source model.

This heat source model is defined by geometrical parameters r_e, r_i, y_e and y_i according to Figure 5 together with the laser power P and laser efficiency η [26]. The efficiency of the laser heat source was supposed to be 75 %. The following values of parameters were applied in this simulation: $r_e = 0.79$ mm, $r_i = 0.7$ mm, $y_e = 2$ mm and $y_i = 0$ mm.

In the static analysis, the symmetry boundary condition was defined on the area in the centre of the weld joint. In accordance with supposed location of welded samples in a fixture, the displacements in y direction in the nodes on the bottom red line (Figure 1) were prescribed as $u_y = 0$. In the node P in the middle of the sample length, all degrees of freedom were removed.

3. Results and discussion

Applying simulation model described in section 2, eighteen numerical experiments (simulations) were performed using the laser welding parameters summarized in Table 2.

Table 2. Welding parameters considered in simulations and the maximum weld pool temperatures.

Simulation No.	P [W]	w [mm.s^{-1}]	T_{max} [°C]
1		30	2494
2		40	2108
3	1600	50	1824
4		60	1610
5		70	1484
6		30	2764
7		40	2341
8	1800	50	2027
9		60	1791
10		70	1607
11		80	1492
12		40	2571
13		50	2231
14		60	1980
15	2000	70	1767
16		80	1604
17		90	1496
18		100	1463

Figure 6 illustrates the temperature fields in chosen times during the laser beam welding of CMnSiNb sheets with the laser power of 2000 W and the welding speed of 70 mm.s^{-1} (simulation No. 15). The fusion zone is very narrow and the temperatures of the base metal distant from the weld centreline remain practically unchanged.

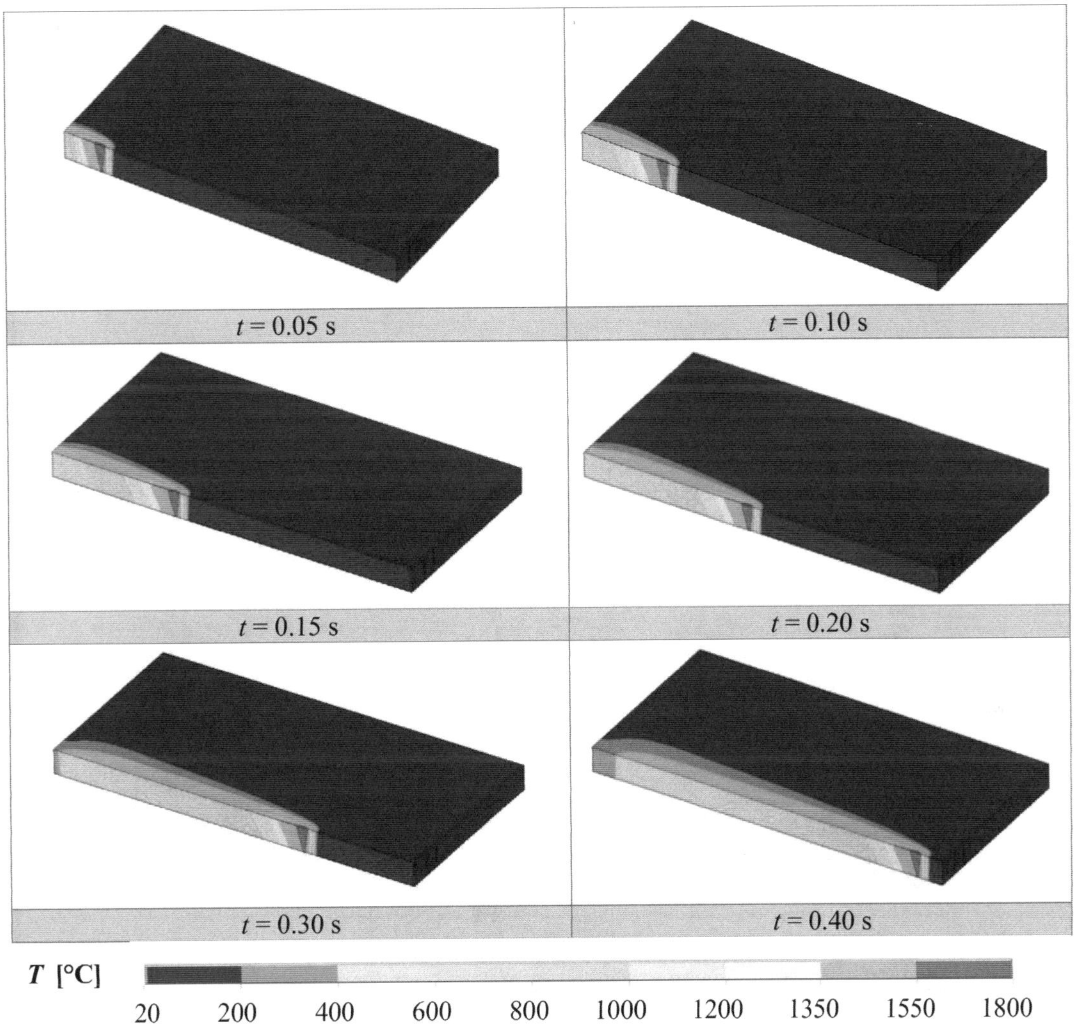

Figure 6. Temperature distribution in chosen times during the laser welding process (simulation No. 15: laser power of 2000 W, welding speed of 70 mm.s^{-1}).

In the first step, the influence of welding parameters on the maximum temperature of a weld pool was evaluated. The computed maximum temperatures of a weld pool are summarized in Table 2 and graphically presented in the dependence on welding speed for different values of laser power in Figure 7. Based on this type of results, those combinations of welding parameters (laser power & welding speed) that would lead to the formation of poor quality weld joints can be excluded. If the temperatures of a weld pool are too low, the material melting in the zone of weld joint is insufficient. On the other hand, at too high weld pool temperatures undesirable material evaporation can occur. In this reason, the further results of simulations No. 1, 6, 7, 11, 12, 13 and 18 were not processed.

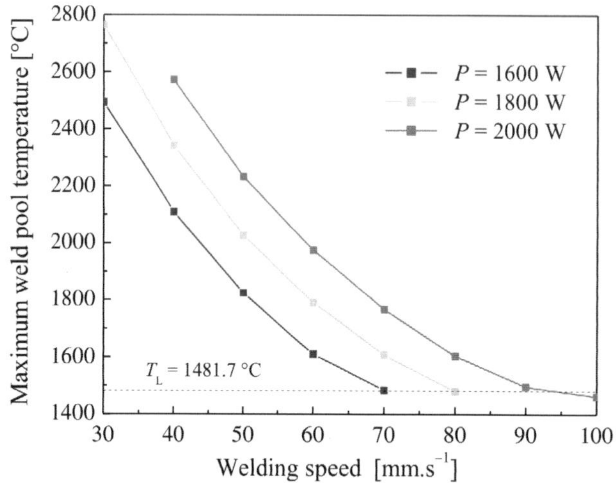

Figure 7. Dependence of the maximum weld pool temperature on the welding speed for considered values of laser power.

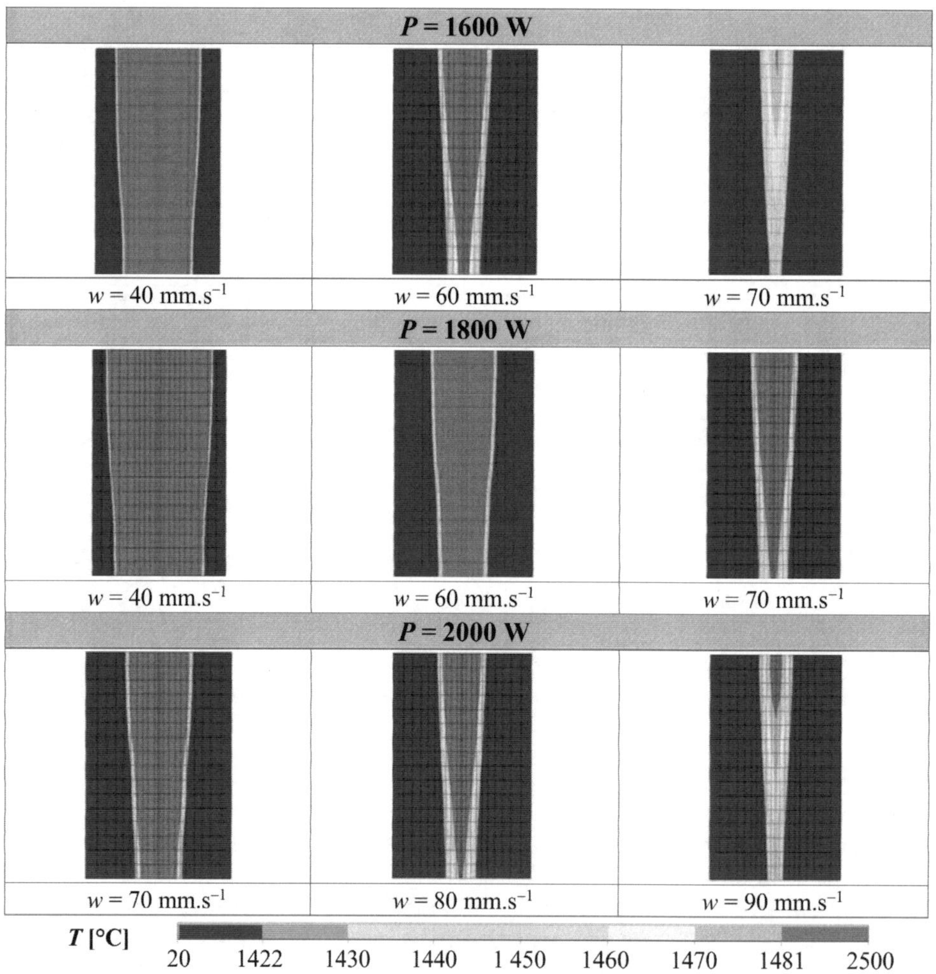

Figure 8. Temperature fields in transversal sections of weld joints for chosen welding parameters.

The width of fusion zone can be proposed according to the temperature fields in the transversal sections of weld joints. Figure 8 shows the temperature fields in the transversal sections of weld joints computed using chosen welding parameters, laser powers and welding speeds. Application of a higher laser power at the same welding speed results in the increased width of fusion zone. The combination of parameters: the laser power of 1600 W and welding speed of 70 mm.s^{-1} or the laser power of 2000 W and welding speed of 90 mm.s^{-1} cannot lead to the production of a sound weld joint as the weld root is not molten.

In Figure 9, the temperature distribution and shapes of weld pools for the welding speed of 70 mm.s^{-1} and different values of laser power are depicted. In gray zones inside the weld pool, the temperatures are higher than the liquidus temperature of the CMnSiNb steel (T_L = 1481.7 °C) and the material is fully melted. Based on this type of simulation results, the dimensions of fusion zones at different welding conditions were predicted. The obtained dimensions of welding pool are summarized in Table 3.

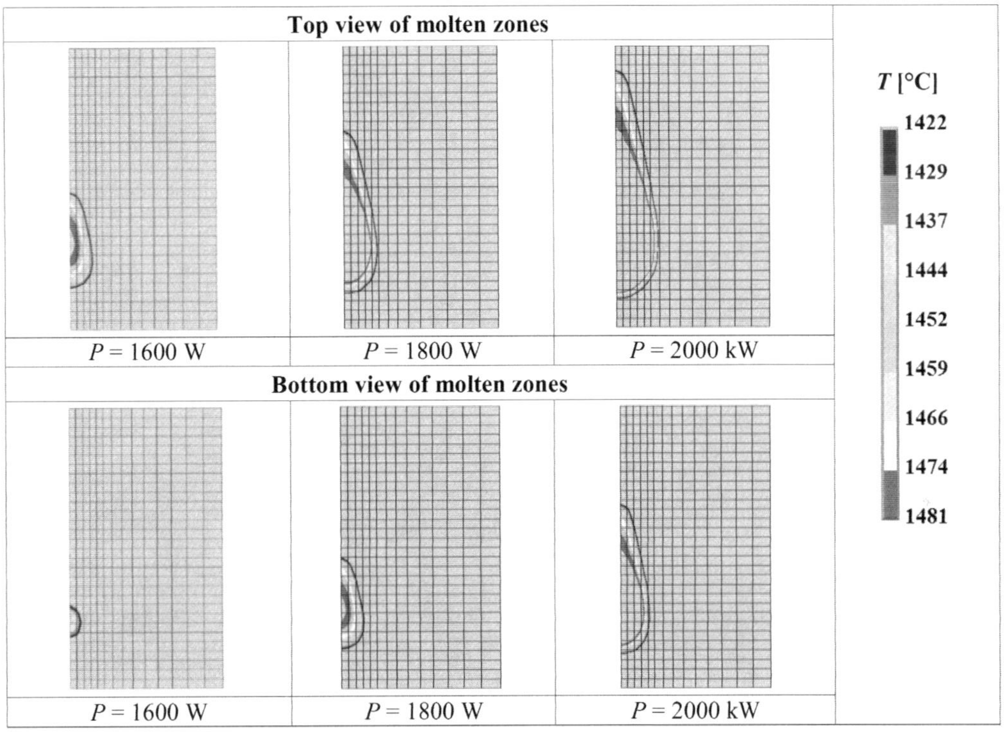

Figure 9. Top and bottom view of fusion zones at welding speed of 70 mm.s^{-1} and different values of laser power.

Table 3. Dimensions of fusion zones for the different combinations of laser power and welding speed.

Simulation No.	*P* [W]	*w* [mm.s^{-1}]	Top width [mm]	Bottom width [mm]	Top length [mm]	Bottom length [mm]
3	1600	50	0.686	0.428	1.895	1.142
4		60	0.412	0.108	1.136	0.333
8	1800	50	0.774	0.562	2.419	1.675
9		60	0.576	0.384	1.773	1.073
10		70	0.394	0.066	1.148	0.283
14	2000	60	0.754	0.572	2.590	1.739
15		70	0.576	0.364	1.784	1.084
16		80	0.364	0.056	1.178	0.262
17		90	0.138	0	0.536	0

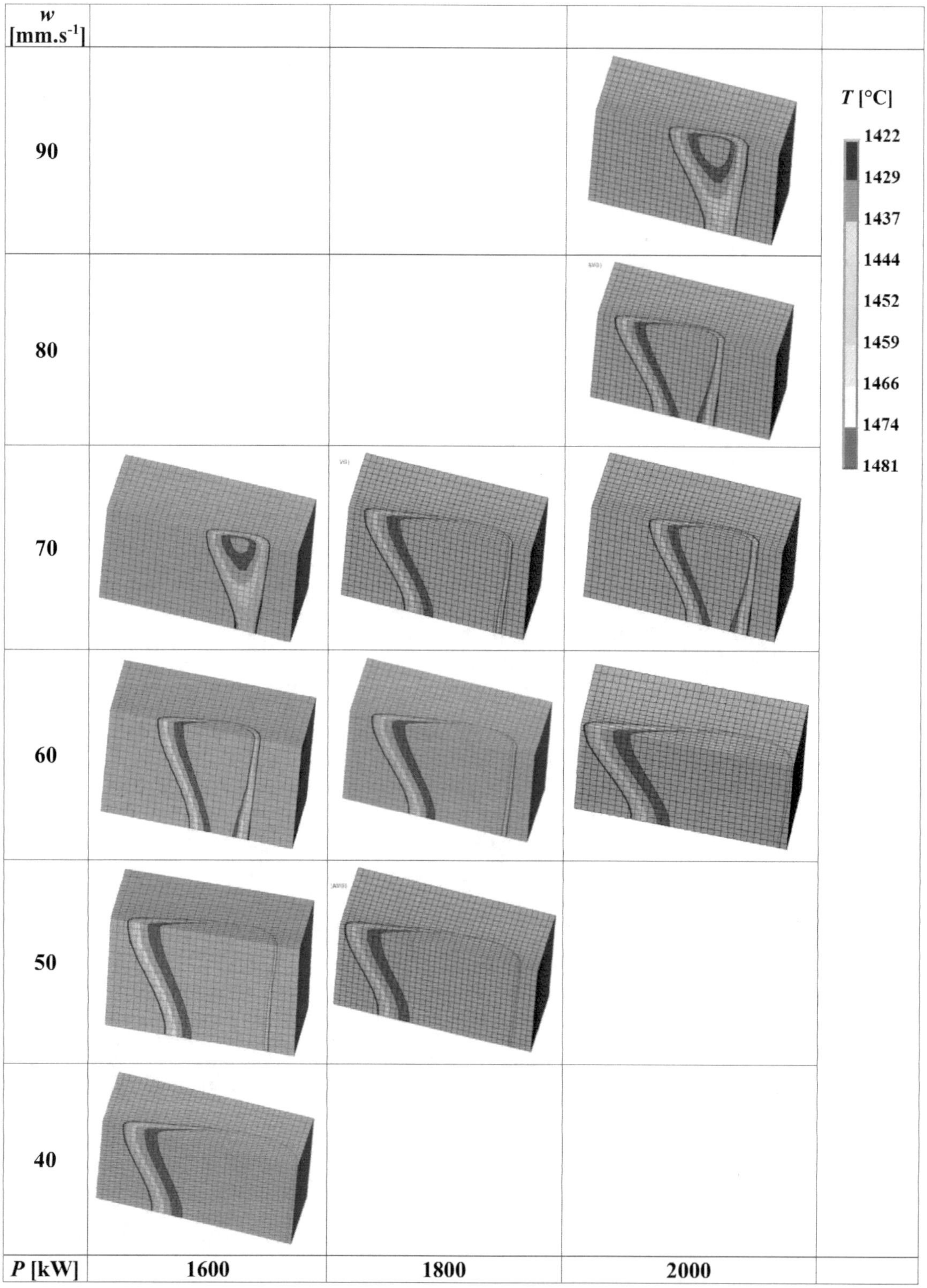

Figure 10. 3D view of fusion zones for analysed combinations of laser power and welding speed.

Figure 10 provides the 3D views of computed dimensions and shapes of welding pools for analysed laser power and welding speed values. Based on the obtained results of thermal analyses, the following combinations of welding parameters can be recommended for the laser welding of 2 mm thick sheets of CMnSiNb TRIP steel:

1. welding speed of 50 mm.s^{-1} and laser power of 1600 W,
2. welding speed of 60 mm.s^{-1} and laser power of 1800 W or
3. welding speed of 70 mm.s^{-1} at the laser power of 2000 W.

In terms of the required high productivity and production efficiency in the automotive industry, the laser welding with the highest welding speed and the use of the maximum power of applied TRUMPF TruDisk 4002 disc laser is expected.

For these welding parameters corresponding to the simulation No. 15, the static analysis was carried out. The distribution of equivalent von Mises stresses are shown in Figure 11 along with the details of stress fields in the weld longitudinal and transversal sections. The maximum von Mises stresses reach the values at the level of 665 MPa in front of the fusion zone. According to experimental results [22], the ultimate tensile strength of welded CMnSiNb TRIP steel depends on the parameters of its thermo-mechanical treatment, achieving the values more then 800 MPa.

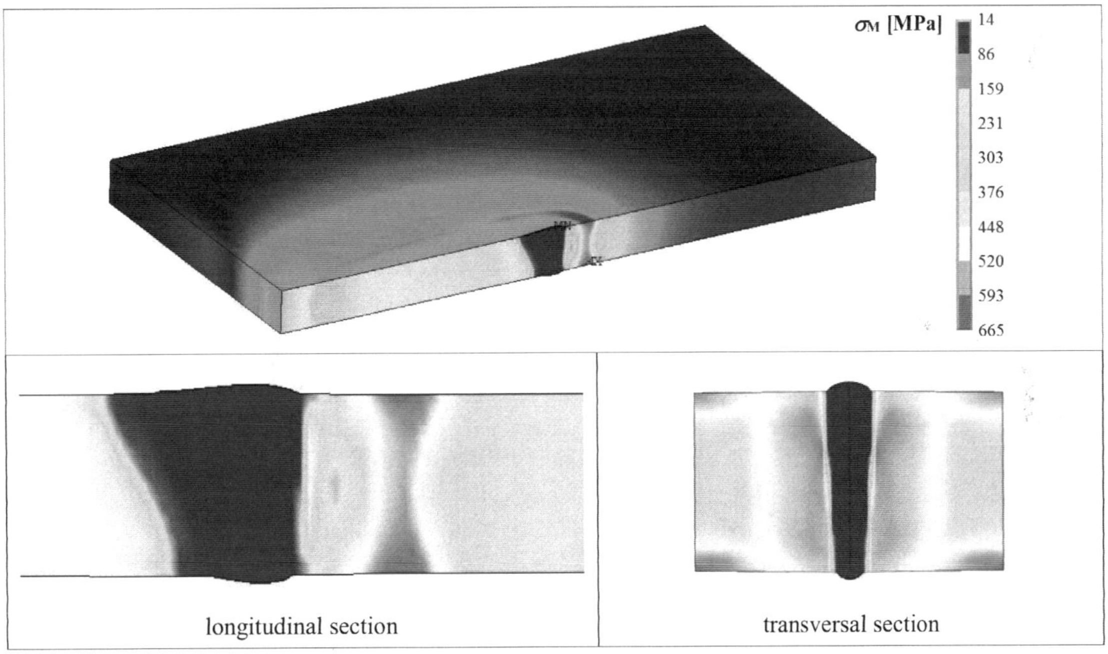

Figure 11. Equivalent von Mises stress computed during the laser welding at the welding speed of 70 mm.s^{-1} and the laser power of 2000 W with details in the longitudinal and transversal sections.

4. Conclusions

The paper deals with numerical simulation of the laser beam welding of CMnSiNb TRIP steel sheets with the thickness of 2 mm with the aim to predict the temperature and stress-strain fields arising during the welding process.

Using the developed simulation model, the effect of welding parameters including the welding speed and the laser power on the shape and dimensions of fusion zone was analysed. Based on the obtained results of thermal and static analyses, the proper welding parameters for production of butt joints of CMnSiNb TRIP steel sheets using the TruDisk 4002 disc laser with the maximum power of 2 kW were designed. The indicative operating diagram, laser power *vs.* welding speed, was prepared. This diagram cotains recommended combinations of welding parameters covering the values of laser power from 1600 W to 2000 W and the welding speeds from the range of 50 mm.s^{-1} to 70 mm.s^{-1}.

To verify the developed simulation model and results of numerical simulations, the experimental butt weld joints of CMnSiNb TRIP steel sheets will be prepared applying the laser beam welding. The weld quality will be evaluated using microstructural analyses and mechanical testing.

Acknowledgment
The research has been supported by the Scientific Grant Agency of the Slovak Republic within the project VEGA No. 1/1010/16 and KEGA No. 029STU-4/2018.

References
[1] Keller S, Kimchi M and Mooney P J 2017 *Advanced High-Strenght Steels Aplication Guidelines Version 6.0.* WorldAutoSteel, April 2017 Available at: https://www.worldautosteel.org/projects/advanced-high-strength-steel-application-guidelines/ (Accessed 7.5.2018)
[2] Schmitt J-H and Iung T 2018 New developments of advanced hight – strength steels for automotive applications *C. R. Physique* **19** 641–656
[3] Fonstein N 2015 *TRIP Steels. In: Advanced High Strength Sheet Steels* (Cham: Springer)
[4] Takahashi M 2003 *Development of High Strength Steels for Automobiles - Nippon Steel Technical Report.* Available at: https://www.nipponsteel.com/en/tech/report/nsc/pdf/n8802.pdf (Accessed 10.5.2019)
[5] Jirkova H, Masek B, Stadler C and Jenicek S 2016 I Capabilities of Unconventional Processing of Multiphase AHSS Steels *IOP Conf. Ser.: Mater. Sci. Eng.* **118** 012023
[6] Galán J. et. al 2012 Advanced high strength steels for automotive industry *Rev. Metal.* **48** (2) 118-131
[7] Li L *et al 2007* Design of TRIP Steel With High Welding and Galvanizing Performance in Light of Thermodynamics and Kinetics *J. Iron Steel Res. Int.* **14** 37-41
[8] Kah P, Pirinen M, Suoranta R and Martikainen J 2014 Welding of Ultra High Strength Steels *Adv. Mat. Res.* **849** 357-365
[9] Brauser S. et. al 2010 Deformation behaviour of spot-welded high strength steels for automotive applications *Mater. Sci. Eng. A* **527** 7099-7108
[10] Hilditch B. T. et. al 2007 Effect of susceptibility to interfacial fracture on fatigue properties of spot-welded high strength sheet steel *Mater. Design* **28** 2566-2576
[11] Rajashekhar S. Sharma, Molian P. 2009 Yb:YAG laser welding of TRIP780 steel with dual phase and mild steels for use in tailor welded blanks *Mater. Design* **30** 4146-4155
[12] Lun N et. al 2017 Microstructure and mechanical properties of fibre laser welded medium manganese TRIP steel *Mater. Design* **131** 450–459
[13] Grajcar A. et. al 2017 Microstructural aspects of bifocal laser welding of trip steels *Arch. Metall. Mater.* **62** 611-618
[14] Evin E and Tomáš M 2017 The Influence of Laser Welding on the Mechanical Properties of Dual Phase and Trip Steels *Metals* **7** 239
[15] Pekovič M e al. 2018 Welding of 42SiCr high-strength steel *Manuf. Technol.* **181** 84-89
[16] Švec P, Schrek A and Dománková M Microstructure evolution of fibre laser welded TRIP and HSLA steel sheets with different thicknesses *Kovove Mater.* **57** 219-227
[17] ANSYS® Academic Research Mechanical, Release 18.1
[18] Incropera P F and De Witt D P 1996 *Fundamentals of heat and mass transfer* (New York: J. Wiley and Sons)
[19] Papatriantafillou I, Agoras, M, Aravas N and Haidemenopoulos G 2006 Constitutive modeling and finite element methods for TRIP steels *Comput. Methods Appl. Mech. Eng.* **195** 5094-5114
[20] Wróbel J, Kulawik A and Bokota A 2017 The numerical analysis of the hardening phenomena of the hot-work tool steel *Procedia Eng.* **177** 33-40

[21] Ronda J and Oliver G. J. 2000 Consistent thermo-mechano-metallurgical model of welded steel with unified approach to derivation of phase evolution laws and transformation-induced plasticity *Comput. Methods Appl. Mech. Eng.* **189** 361-417

[22] Kučerová L and Bystrianský M Comparison of thermo-mechanical treatment of C-Mn-Si-Nb and C-Mn-Si-Al-Nb TRIP steels *Procedia Eng.* **207** 1856-1861

[23] JMatPro Help, Release 6.1 2012 Sente Software Ltd

[24] Huy H and Argyropoulos S A 1996 Mathematical modelling of solidification and melting: a review *Model. Simul. Mater. Sci.* **4** 371-396

[25] Podešva P (2019) *Application of the ANSYS software for the design of welding parameters for laser welding of high-strength steels* (Diploma Thesis) Slovak University of Technology in Bratislava (MTF SUT)

[26] Wu C S, Wang G, Zhang Y M 2006 A new heat source model for keyhole plasma arc welding in FEM analysis of the temperature profile *Weld J* **85** 284-291

Prediction and modeling of roughness in ball end milling with tool-surface inclination

O Bilek, R Milde, J Strnad, M Zaludek and M Bednarik

Tomas Bata University in Zlín, Faculty of Technology, Vavreckova 275, Zlin, Czech Republic

bilek@utb.cz

Abstract. The quality of mill finishing of shaped surfaces is influenced by a number of input parameters. Current automated manufacturing systems allow adaptation of the machining process aiming at the final surface quality. Despite all the advantages, these systems require a behavioural model, a prediction of the output, based on the input parameters. Some of these models are summarized in this paper, including contemporary evaluated models as well as their functional dependencies; moreover, offers an application of mill finishing with a ball-end cutter incorporating tool axis or surface inclination.

1. Introduction

Milling is a conventional technology in which the material is removed from the workpiece in the form of chips with a multi-blade cutting tool. The main rotary movement is performed by the tool and the chips are periodically generated according to the number of tool blades. The resulting surface quality of the machined surface consists of a series of elementary surfaces formed by individual cutting edge as well as by entire cutting process. Surface topography is the result of the material removal process due to the relative movement of the tool and the workpiece. In the primary view, it is possible to find a direct correlation of the resulting surface quality with the geometric characteristics of the milling tool.

Predicting surface quality after milling is a challenge at many levels. This is because the surface quality and the associated evaluation parameters are influenced by a number of input factors whose behavior is often non-linear, moreover some of the factors interact with each other. In addition to the primary concern, surface quality is important to the performance of a part and affects, inter alia, durability, friction properties, corrosion resistance and lubricant distribution, heat transfer, light reflectance and fatigue strength, and last but not least the appearance and cost of the part [1-6].

Surface quality is a widespread product quality evaluation parameter and is one of the technological requirements. The following article discusses the surface quality after milling, which is used in multi-axis finish milling. A spherical milling tool is used when milling contoured surfaces with a slope. This avoids the known phenomenon where, in the vicinity of the tool axis, the resulting cutting speed approximate to zero and at this point the material is not being cut.

In the industry, especially in the manufacture of forming tools, mold cavities for foundry and injection molds in particular, surface quality is characterized as the arithmetical mean roughness *Ra*. The *Ra* parameter is widely used in manufacturing practice, although it provides limited information on the condition and quality of the machined surface. This parameter is most often accompanied by the measurement of the parameter *Rz*, however dependent on *Ra*.

2. Prediction models and optimization techniques

Despite the fact that the machining process is influenced by a number of factors related to the machine-tool-workpiece-fixture relationship [7-9], current process machining models are able to suggest the setting of inputs, most often cutting conditions, according to the required surface quality [10]. The methodology of surface quality prediction includes various solution approaches; from simpler ones such as the determination of the kinematic model of surface creation, experimental investigation and surface analysis, the implementation of artificial intelligence (AI) as well as approaches using the design of experiment. Despite a number of methods and models, the researches almost without exception describe surface quality only for a specific case with given boundary conditions. In the absence of a universal surface quality model, re-determination of the input change model is required. A disadvantage of many surface quality models is also the fact that the cutting conditions proposed can be given outside the machine's working range or require precise adjustment sensitivity. Recent studies have focused on investigating the cutting process of the end mill with spherical tool end, known also as a ball end mill cutter, for finishing oblique and contour shaped surfaces [11-12]. Yet few researchers are involved in this issue and there are not enough surface quality models suitable for application in manufacturing practice.

Several important studies are devoted to the relationship between surface quality and the machining process. The authors Benardos and Vosniakos [13] not only determined graphically the factors influencing the cutting process, their great contribution in the area of prediction is the summary and classification of solution approaches. Another general solution is proposed by Lu [14] in the study of the prediction of surface roughness using the artificial intelligence with the contribution of vibration. Quintana [15-19] also continually addresses the relationship between ball milling, surface quality and process stability. Based on previous research in this area, it is possible to categorize the surface quality prediction into the following groups:

a) Kinematic models (based on machining theory)
b) Empirical modeling (experimental analysis of effects, regression analysis)
c) Planned experiment methods (DOE)
d) Artificial intelligence methods (ANFIS, ANN)
e) Advanced methods (GA, Fuzzy, Neuro-Fuzzy, Hybrid methods)

Sadilek and Cep [19] present one of the mechanical *Ra* model (1) for planar surfaces. They emphasize the need to mill off-axis due to zero cutting speed near the tool axis. Therefore, they recommend tilting the tool or workpiece surface and using multi-axis machining operations. Further mechanical model (2) gives Peterka [20], where substitution is performed for inclined tool-workpiece machining, allowing the calculation of theoretical *Ra* values. The model is verified experimentally for various machining methods.

$$Ra = \frac{R^2}{a_e} \times \left\{ \begin{array}{l} arc2\left[\arccos\left(\frac{1}{2}\cos\arcsin\frac{a_e}{2R} + \frac{R}{a_e} arc\arcsin\frac{a_e}{2R}\right)\right] - \\ -\sin 2\left[\arccos\left(\frac{1}{2}\cos\arcsin\frac{a_e}{2R} + \frac{R}{a_e} arc\arcsin\frac{a_e}{2R}\right)\right] \end{array} \right\} \times 1000 \quad (1)$$

$$Ra = \frac{R^2 \cos\alpha}{a_e} \left\{ \begin{array}{l} arc2\left[\cos^{-1}\left(\frac{1}{2}\cos\sin^{-1}\frac{a_e}{2R\cos\alpha} + \frac{R\cos\alpha}{a_e}\arcsin^{-1}\frac{a_e}{2R\cos\alpha}\right)\right] - \\ -\sin 2\left[\cos^{-1}\left(\frac{1}{2}\cos\sin^{-1}\frac{a_e}{2R\cos\alpha} + \frac{R\cos\alpha}{a_e}\arcsin^{-1}\frac{a_e}{2R\cos\alpha}\right)\right] \end{array} \right\} \times 1000 \quad (2)$$

An important author in this area is Quintana, who presents mechanical *Ra* models for planar and inclined surfaces. The results show a relatively good agreement of the theoretical *Ra* with the measurement, but the theoretical calculation of the height inequality in an independent equation, in some cases, varies by up to 75 %. The author explains this mismatch by the thermal contribution and the dynamic effect of the milling process.

Quintana suggests monitoring the process using cause-effect methods using ANN, GA, Fuzzy, or expert systems. The artificial neural networks are used in the later study [15], and focuses on *Ra* modeling for planar surfaces. These relationships and equation (3)

$$Ra = \frac{2}{a_e}\left(R^2 \cos^{-1}\left(\frac{\sqrt{R^2 - \frac{a_e^2}{4}}}{2R} \right) + \frac{R}{a_e}\sin^{-1}\left(\frac{a_e}{2R}\right) \right) - \left(\frac{R^2}{a_e}\sin^{-1}\left(\frac{a_e}{2R}\right) + \frac{\sqrt{R^2 - \frac{a_e^2}{4}}}{2} \right) \times$$
$$\times \left[\frac{2R^3}{a_e}\sin^{-1}\left(\frac{a_e}{2R}\right) + 2R^2 - R\sqrt{R^2 - \frac{a_e^2}{4}} - \left(\frac{R^2}{a_e}\sin^{-1}\left(\frac{a_e}{2R}\right) + R - \frac{\sqrt{R^2 - \frac{a_e^2}{4}}}{2} \right)^2 \right]^{1/2} \quad (3)$$

are inputs for ANN system using a dual-layer feed-forward network, a Levenberg-Marquardt learning algorithm and one hidden layer of 20 neurons. The entire model includes, among other inputs, vibration, the effect of cooling, and in total Quintana conducted 250 experiments. The obtained excellent degree of accuracy between the model and the measured data shows that ANN is the right way for modeling the machining process.

Suresh [21] offers two mathematical models of *Ra*, which were compiled from measured data and evaluated by RSM. However, the first order model in this paper offers limited descriptiveness, and therefore a second order polynomial model (4) with a regression coefficient $R^2 = 0.9305$ is established. The equation is valid for milling planar surfaces with side radius cutters. The author used genetic algorithms with a population size of 20 and a maximum number of 500 generations to find the optimal cutting geometry of the tool and cutting conditions for the best surface quality.

$$Ra = 0.20924 - 0.0574(\ln v_c + 0.793536)^2 + 0.0215(\ln v_f + 3.5907)^2 + \\ + 0.3506(\ln \gamma + 0.4066)^2 + 0.2746(\ln r_\varepsilon + 0.394756)^2 \quad (4)$$

The Jatti [22] model (5) uses RSM with the Box Behnken method with three-level factorial design and other centered, which is a 2nd order polynomic function and serves as input for GA optimization.

$$Ra = 9.01938 + 0.00194971n - 0.0101682v_f - 6.41036a_p - 2.54626\times10^{-7}n^2 + \\ + 4.08583\times10^{-6}v_f^{\,2} + 5.92333a_p^{\,2} - 3.42857\times10^{-7}n\times v_f - \\ - 4.31429\times10^{-4}n\times a_p - 2\times10^{-4}v_f\times a_p \quad (5)$$

The model works with an average error of 5.9 %, a maximum error of 12.84 % for machining Al-Si12 alloys, used for specific application in foundry purpose.

Explanation of parameters and abbreviations in equations (1-5) and chapter above is given as:

R	cutting tool radius (mm)
v_c	cutting speed (m/min)
n	revolutions per minute (min^{-1})
v_f	feed rate (mm/min)
f_z	feed per tooth (mm)
a_p	axial depth of cut (mm)
a_e	radial depth of cut (mm)
r_ε	mill cutter radius (mm)
α	surface / tool inclination angle (°)
γ	tool rake angle (°)
RSM	Response surface methodology
GA	Genetic algorithm
ANN	Artificial neural networks
DOE	Design of experiments

3. Influence of inclination angle

Shape surfaces on parts can be classified into two categories for the purpose of this study. Shallow surfaces are characterized by the inclination angle α relative to the horizontal surface (0 ° - 30 °>, the steep surfaces then within the inclination angle α (30 ° - 90 °).

It can be seen from the evaluation and in the figure 1 that, when milling a surface with a slight slope, referred to as shallow, the better surface quality is achieved under the same technological conditions than in the case of surfaces with a greater steepness. The trend of the dependence on the angle of inclination is exponential with slight gradient. The behavior is corresponding to the variating cutting forces as the angle of inclination changes, the low stiffness of the tool in the lateral directions, vibrations and other dynamic machining phenomena.

The radial depth of cut a_e is identified by literature and analytical analysis as the most significant influencing parameter on surface quality when milling with a ball end mill tool. If the parameter a_e is rising, which means that the spacing between parallel cuts increases, this obviously leads to a deterioration of the surface quality.

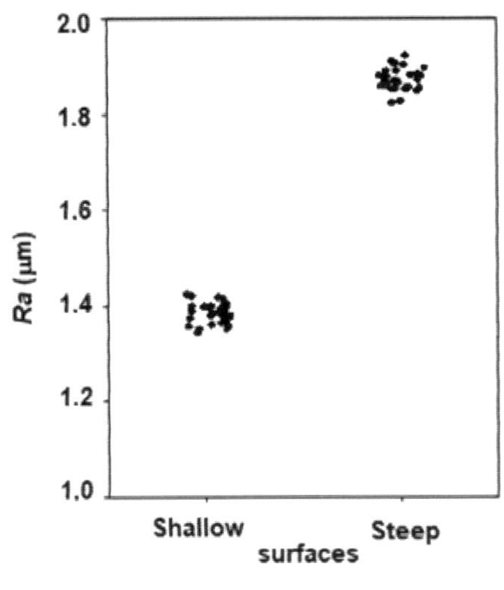

Figure 1. Individual *Ra* sets of milling surfaces with shallow and steep inclination angle ($\alpha_{shallow}$ = 15 %, α_{steep} = 75 %, a_e = 0.6 mm, f_z = 0.1 mm).

As can be seen in the figure 2, the course of the roughness parameter *Ra* is exponential, and mathematical interpretation can evaluate recursively values that cannot be measured, moreover predict the *Ra* roughness for specific machining conditions. A frequent limiting criterion is the time the tool is engaged into the cut. The parameter is directly related to tool life and tool wear; at the same time, it is associated with the cost of operating the machine for machining large-dimension parts and allows the selection of process parameters with respect to time constraints.

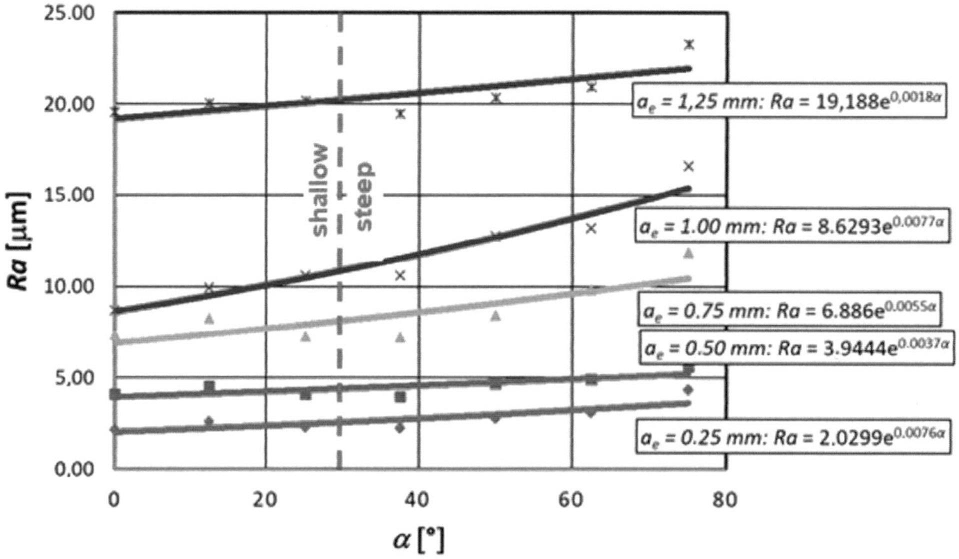

Figure 2. Dependence of *Ra* on the inclination angle α and the radial depth of cut a_e (AW 7022, $D = 5$ mm).

4. DOE and significant parameters

In general, design of experiments (DOE) is used to obtain process information based on a methodological plan. Not all possible combinations of options need to be measured to determine process parameters, saving time and resources. On the other hand, it requires precious accuracy when determining output parameters with minimal variance.

With respect to the mentioned facts, the data were evaluated during milling of inclined surfaces. DOE allows to analyze multiple input parameters in one cycle. Based on the experiments and their values transformed into the signal-to-noise ratio, the main effects (table 1) and their order of significance by factors were determined.

Table 1. Order of effects by factor.

Level	1	2	3	4	5
***F*1**	α	D	l	a	v_f
***F*2**	a	D	l	v_f	α
***F*3**	D	a	v_f	l	α
Ra	a	v_f	l	D	α
Rz	a	v_f	l	D	α
Rmr	D	a	v_f	α	l

*F*1	passive force (N)
*F*2	feed force (N)
*F*3	main cutting force (N)
Ra	arithmetical mean roughness value (µm)
Rz	mean roughness depth (µm)
Rmr	material component of the profile (µm)
α	surface / tool inclination angle (°)
D	cutting tool diameter (mm)
a	radial depth of cut (% × ∅*D*)
l	cutting tool ejection in the clamping system (mm)
v_f	feed rate (mm/min)

According to the first two levels, the cutting force components are influenced primarily by the tool selection and its diameter *D*, the radial depth of cut *a* and the angle of inclination α in one case. The main cutting force *F3* is the largest component to the total force resultant and its size is influenced by the tool diameter *D*, while the smallest contribution has the inclination angle α, which for the finish milling relates to the small size chip cross-section. The feed force *F2* is influenced mainly by the radial depth of cut *a*, and similarly to the *F3* inclination angle does not play significant role. On the other hand, the magnitude of the passive force component *F*1 is affected by the tool geometry and inclination angle α, however the smallest influence has the feed rate.

The roughness height parameters *Ra*, *Rz* are interdependent parameters that are determined from the same roughness profile. Therefore, they have the same order of significance as can be seen in the table 1. The material proportion *Rmr* is significantly influenced by the tool (mill cutter diameter *D*) and radial depth of cut *a*. These are primarily variables that affect the frequency characteristics of the roughness profile and can be said to have particular influence on the regularity of movement during machining and creation of the new surface. The experiment revealed that the extrusion of the tool and the associated stiffness of the tool have no significant effect on the material component of the profile.

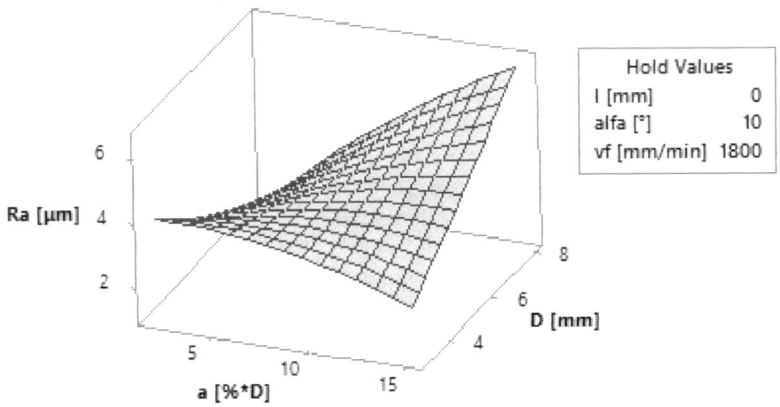

Figure 3. Surface plot of *Ra* for radial depth of cut *a* and tool diameter *D*.

The most ruling milling parameters for optimum surface quality are radial depth of cut *a*, and tool geometry, characterized by the diameter *D* of the mill cutter. In the plot of figure 3, a deterioration of the surface quality with increasing parameters *a* and *D* can be observed. However, it is necessary to note instability at smaller radial depths and the for small tool diameters, where surface quality is influenced by other associated process phenomena, resulting in an improvement in the observed parameter *Ra*. The overall behavioral model obtained by the DOE has a mathematical interpretation in the following equation (6):

$$\begin{aligned} Ra = &-132.3 + 1.186 \times l - 0.2673 \times a - 1.12 \times D + 0.2034 \times \alpha + \\ &+ 0.1537 \times v_f - 0.02027 \times l \times l - 0.009731 \times a \times a + 0.031 \times D \times D - \\ &- 0.007986 \times \alpha \times \alpha - 0.00043 \times v_f \times v_f + 0.008198 \times l \times a - \\ &- 0.02412 \times l \times D + 0.000322 \times l \times \alpha - 0.000455 \times l \times v_f + \\ &+ 0.1055 \times a \times D \end{aligned} \tag{6}$$

The obtained equation shows the dependence of the parameter *Ra* on the cutting conditions (a, v_f), the tool geometry (*D*) but also on the length of the tool ejection *l* above the recommended settings.

5. Conclusion

Only a small percentage of research is contributed to modeling, predicting, and optimizing the surface quality of inclined and shaped surfaces after machining. On the basis of the research analysis it can be observed that the surface quality prediction of many theoretical models include a significant error between the measured and predicted values or the application is possible only in a limited range of materials, tools and cutting conditions. Despite this fact, we can observe the increasing demand for the new methods of processing engineering data using advanced statistics, design of experiments and artificial intelligence.

Machining technology by milling does not lose its importance even as finishing operation. The mill finishing may achieve comparable close surface quality to grinding with commercially available tools and machine options. The aim is to minimize or prevent subsequent finishing of the surfaces by grinding, which arise undesirable temperature field and stress states into the surface layer.

Acknowledgements

This work and the project was realized with the financial support of the internal grant of TBU in Zlín No. IGA/FT/2019/001 funded from the resources of specific university research.

References

[1] Mikó B, Tóth B and Varga B 2017 Comparison of Theoretical and Real Surface Roughness in Case of Ball-End Milling *Solid State Phenom* **261** 299–304

[2] Stanek M, Manas M, Ovsik M, Reznicek M, Senkerik V and Fluxa P 2018 Surface quality of injection molds *Int. J. Mech.* **12** 246–51

[3] Matras A and Kowalczyk R 2014 Analysis of machining accuracy during free form surface milling simulation for different milling strategies *Photonics Applications in Astronomy, Communications, Industry, and High-Energy Physics Experiments* ed R S Romaniuk p 929019

[4] Monkova K and Monka P 2011 Relationship of surface roughness and machining productivity *Annals of DAAAM and Proceedings of the International DAAAM Symposium* (Vienna: Danube Adria Association for Automation and Manufacturing, DAAAM) 193–4

[5] Beno J, Mankova I, Izol P and Vrabel M 2016 An approach to the evaluation of multivariate data during ball end milling free-form surface fragments *Measurement* **84** 7–20

[6] Hricova J and Naprstkova N 2015 Surface roughness optimization in milling aluminium alloy by using the Taguchi's design of experiment *Manuf. Technol.* **15** 541–6

[7] Balázs M. 2015 Study on tolerance of shape and orientation in case of shoulder milling *Development in machining technology* ed W. Zebala **5** (Cracow: Cracow University of Technology) 136–50

[8] Felhő C, Karpuschewski B and Kundrák J 2015 Surface Roughness Modelling in Face Milling *Procedia* CIRP **31** 136–41

[9] Zebala W 2016 The Influence of Tool Stiffness on the Dimensional Accuracy in Titanium Alloy Milling *Key Eng Mater* **686** 108–13

[10] Buj-Corral I, Vivancos-Calvet J and Domínguez-Fernández A 2012 Surface topography in ball-end milling processes as a function of feed per tooth and radial depth of cut *Int. J. Mach Tools Manuf.* **53** 151–9

[11] De Souza AF, Diniz AE, Rodrigues AR and Coelho RT 2014 Investigating the cutting phenomena in free-form milling using a ball-end cutting tool for die and mold manufacturing *Int. J. Adv. Manuf. Technol.* **71** 1565–77

[12] Izol P, Fabian M, Kopas M, Fedorko G and Fabianova J 2014 Evaluation of surface milling strategies using selected elements of machined shapes of forging die cavities *Manuf. Technol.* **14** 178–86

[13] Benardos PG and Vosniakos GC 2003 Predicting surface roughness in machining: A review *Int J. Mach. Tools Manuf.* **43** 833–44

[14] Lu C 2008 Study on prediction of surface quality in machining process *J. Mater. Process. Technol.* **205** 439–50
[15] Quintana G, Garcia-Romeu ML and Ciurana J 2011 Surface roughness monitoring application based on artificial neural networks for ball-end milling operations *J. Intell. Manuf.* **22** 607–17
[16] Quintana G, Bustillo A and Ciurana J 2012 Prediction, monitoring and control of surface roughness in high-torque milling machine operations *Int. J. Comput. Integr. Manuf.* **25** 1129–38
[17] Quintana G, Ciurana J De and Ribatallada J 2010 Surface Roughness Generation and Material Removal Rate in Ball End Milling Operations *Mater. Manuf. Process* **25** 386–98
[18] Quintana G, Rudolf T, Ciurana J and Brecher C 2011 Surface roughness prediction through internal kernel information and external accelerometers using artificial neural networks *J. Mec.h Sci. Technol.* **25** 2877–86
[19] Sadílek M, Čep R, Budak I and Soković M 2011 Aspects of using tool axis inclination angle *Stroj. Vestnik/Journal Mech. Eng.* **57** 681–8
[20] Peterka J 2004 A New Approach to Calculating the Arithmetical Mean Deviation of a Profile during Copy Milling *J. Mech. Eng.* **50** 594–7
[21] Suresh Kumar Reddy N and Venkateswara Rao P 2005 A Genetic Algorithmic Approach for Optimization of Surface Roughness Prediction Model in Dry Milling *Mach. Sci. Technol.* **9** 63–84
[22] Jatti VS, Sekhar R and Patil RK 2013 Study of ball nose end milling of LM6 al alloy: Surface roughness optimisation using genetic algorithm *Int. J. Eng. Technol.* **5** 2859–65

IOP Conf. Series: Materials Science and Engineering **726** (2020) 012004 doi:10.1088/1757-899X/726/1/012004

Acoustic emission analysis of hard coatings cracking during indentation test

P Drobný[1], Ľ Čaplovič[1], M Sahul[1], P Babincová[1] and V Koula[2]

[1]Slovak University of Technology in Bratislava, Faculty of Materials Science and Technology in Trnava, Ulica Jána Bottu č. 2781/25, 917 24 Trnava, Slovak Republic
[2]ZD Rpety-Dakel, Ohrobecká 408/3, 142 00 Praha 4, Czech Republic

drobny@vuje.sk, lubomir.caplovic@stuba.sk

Abstract. Hard coatings are widely used in industries where the good behaviours of surface properties play the main role in a life cycle of machine components. The main material characteristic of coatings which determines behaviour of surface is Hardness. In the past, Hardness of materials was considered to be the most important material property, which influences wear properties of coatings as well as value of maximal elastic tension needed for initiation of cracks. However, Adhesion of hard coatings depends on mechanical properties such as Young modulus E, hardness H and fracture toughness R. Initiation/Formation of fractures and their propagation in hard coatings is commonly investigated using indentation tests. Creation of Hard coating crack during indentation is usually detected as a wide frequency AE signal. In this article, the initiation of cracks is investigated using data measured during the indentation. The cracks formation/propagation and spalling in hard coatings during indentation were observed and quantified by using AE. AE frequency analysis was calculated from the signals measured during the indentation and it was evaluated from recorded AE signals.

1. Introduction

Basically, three main types of hard coatings cracks deposited on ductile substrate have been identified:

- circumferential cracks which appear at the periphery of plastic zone,
- channel cracks, which are initiated under big stress caused by contact between indenter and coating,
- radial cracks, which come from the middle of indentation imprint and propagate outside in the form of beams.

If the indentation loading force is high enough and the substrate is plastically deformed, the coatings start losing adhesion with surface and then delamination or spalling can occur. Based on this type of damage, it is inevitable to find an appropriate method for diagnostics of hard coatings cracking. Hard coatings cracking and spalling generate the energy which propagates through the material in the form of elastic waves, the formation of which could be detected by an acoustic emission measuring device during instrumented indentation.

When the type of hard high-performance film is designed, it is of great importance to learn the material properties of coating-substrate system. Due to size limitation of coatings the best method for evaluation of material properties of coating-substrate systems are indentation methods. Indentation is widely used for measuring material parameters of coating-substrate systems and many studies have been done and a lot of approaches have been developed in order to evaluate measured data from indentation and to calculate all the main material characteristic parameters. If indentation loading force is high enough and the substrate is plastically deformed, the coatings start losing adhesion with surface and then subsequently delamination or spalling can occur.

Hard coatings usually have a brittle character, and a fracture failure often occurs in a hard thin film when it is subjected to high stress and it is called cracking. This failure of coating adhesion to the substrate usually results in interfacial debonding. One of the most important parameters of coating which indicates its ability to remain attached to the substrate is Adhesion. At present there is no standard test procedure for calculating adhesion of coating, but some authors published methods based on stress or energy approach [3,8]. Authors used the data from indentation and define fracture toughness of the coating by using calculated parameters from indentation data and they estimate critical force Fc from indentation curve where the coating fracture is represented by pop-in. However, the estimation of critical force from loading curves can be difficult for some coatings.

During indentation the stress activated on the border coat – substrate causes creation of cracks which propagates through the surface. When coating breaks, the cracks initiation and propagation release the energy, that propagates through material and can be measured and quantified with Acoustic Emission device. The analysis of acoustic emission signals provides information of releasing this energy into material. Creation and propagation of crack release high energy than acoustic signal exceeds threshold and this event is recorded by AE device. The critical force Fc when the first crack of coating appears could be determined by analysis of AE data. Then the value can be used for calculations of coating's adhesion parameters.

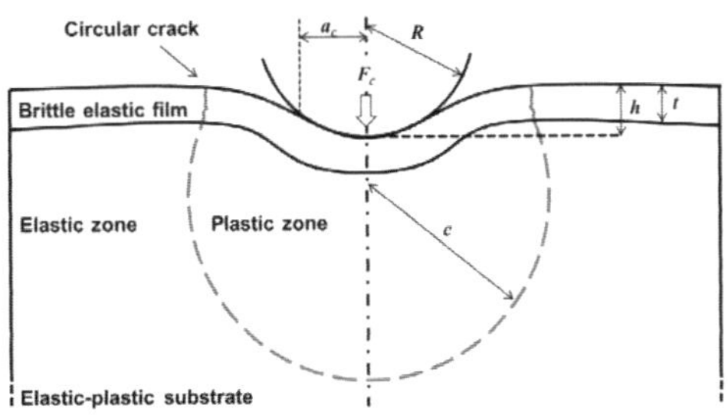

Figure 1. Schematic cross-section of deformation profile of a hard brittle film on an elastic-plastic substrate under indentation [4].

Under contact pressure by indentation, the coating undergoes elastic deformation until brittle fracture. The substrate is deformed plastically at relatively high loads. The first crack of hard coating which occurs under loading by indentation is circumferential crack with the radius c see figure1

When the first crack occurs, indentation force equals to the critical force F_C. Using folowing equation it`s possible to calculate the radius c of the plastic zone in the ductile substrate [7]

$$c = \sqrt{\frac{3F_C}{2\pi\sigma_y}} \quad (1)$$

F_C is the load carried by the substrate at the critical indentation depth at which pop-in occurs. The sign σ_y is the yield stress of the ductile substrate [4,7].

2. Experimental procedure

In this study we investigate acoustic signals generated during indentation test. As an experimental device we used materials testing device Micro-Epsilon UMZ-3k. Indenter Rockwell C was carried out by this device. Acoustic Emission signals released during macroindenation were measured by DAKEL-ZEDO

Device (DAKEL). Surface cracking and spalling of the coatings after indentation test were examined by the JEOL JSM 7600 F high resolution scanning electron microscope (HRSEM).

The studied experimental sample was fabricated using PLATIT π80+DLC PVD technology. Nanocomposite hard coating nACRo based on Cr and silicon were deposited with BIAS 85V on the high-speed steel (HSS 6-5-2-5, STN19 852).

To be able to reach every possible generated acoustic signal the AE device DAKEL was more sensitive during this test what was set up by the increase of the amplification.

3. Macroindentation and AE measuring

The well-known Rockwell C indentation test is prescribed by the VDI 3198 norm, as a destructive quality test for coated compounds [1, 2]. In this study the surface has been loaded with Rockwell C indenter and the test has been executed according Daimler-Benz method (standard VDI 3198) with constantly increasing force at maximum load 50N (instrumented indentation). The applied force through indenter generates firstly elastic and then plastic deformation of sample surface. When coating breaks, the cracks initiation and propagation and spalling of coatings release high energy, which propagates through material and can be measured and quantified by Acoustic Emission device.

Indentation method - „Mercedes test“ is unpretending for execution and is frequently used often when the wear behaviour between coat and substrate is investigated. The stress on the interface coat – substrate is caused by indenter, by static indentation of indenter. The stress activated on the interface coat – substrate causes the creation of cracks which propagates through the surface. The advantage of indentation method is the fast execution of tests, minimal demand on measuring device and possibility of coating behaviour evaluation with acoustic emission monitoring directly on the researched specimens without any further destruction of specimens.

Indenter which is Rockwell cones with the top angle of 120° has been carried by a special device for mechanical behaviour measurement of materials UMZ -3K (Micro-Epsilon). This device can measure the load force in [N] and there is also possibility to adjust the speed of this movement. The Indenter movement stopped automatically when the adjusted force is reached.

Acoustic emission signals generated during the test in the specimens were measured by DAKEL-ZEDO device. DAKEL-ZEDO is modern powerful modular system for detection and evaluation of acoustic emission signals. In this case it`s used for detection of coating cracks and delamination during Mercedes test.

Many authors pointed out that it is possible to identify cracking of coating from the load-displacement curve. During the crack formation, a significant amount of energy is dissipated, which causes the discontinuity in the loading curve [3,4]. This discontinuity is called pop-in event and could be caused by different physical mechanisms. Occurrence of cracks in coating is one of more sources of this discontinuities in the loading curves [5]. In the case that we want to distinguish from the loading curve when a critical Force of first crack occurs could be a real issue. The critical force could be used for calculating of radius c of the plastic zone in the ductile substrate or a fracture toughness of a hard brittle coating on the basis of the fracture pattern caused by a Rockwell indentation of the film/substrate system [6]. Also, it is possible to express if coating is susceptible for cracking by low indentation force.

The maximum reached indentation load was 50 N and has been reached without any delay in loading phase. The 50N load was selected based on the previous experiments confirming that the first acoustic signals caused by the coating breakdown occurred in the range of 10 to 20 N. For better diagnostic of acoustic signals, the amplification was increased to 96 dB. Indentation test parameter are shown in table 1.

Table 1. Indentation test parameters F_{max} = 50 N

Max. force (N)	Holding force (N)	Holding delay (s)	Movement speed (mm/min)	Amplification (dB)
50	-	-	0.2	96

AE parameters used for evaluation were

- Hit - parameter which was registered when AE signal amplitude exceeds adjusted threshold. It indicates number of all cracks and other structural instabilities which were arisen and growth in specimen.
- Hits Energy gives information about energy which is released into material caused by crack initiation and propagation. Bigger crack initiation causes more energy released into material. Bigger amount of energy is usually released when hard coating spalling or delaminating occur.

4. Results and discussion

The problem to find the pop-in in the loading curves is also a point in our experimental macroindentation loading curve. In the figure 2. is load-displacement curve measured by test device UMZ 3K with maximum indentation load 50N. From the load displacement curve isn`t possible to define exactly the localisation of "pop-in" neither the critical load force. It is possible to see a lot of instabilities on the curve but it`s difficult to find out when the first "pop-in" occurred.

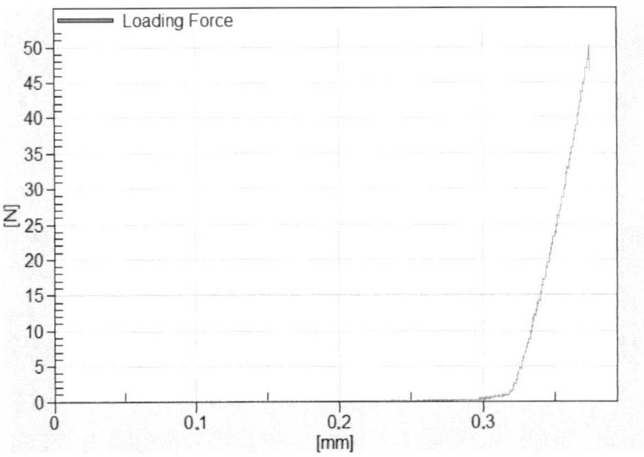

Figure 2. Load-displacement curve, F_{max} = 50 N, sample nACRo with BIAS 85V

For the reason with determination of pop-ins in load-displacement curve, the acoustic signal Hits curve which represents the number of AE events (possible instabilities in the material) recorded during loading phase were added in Figure 3.

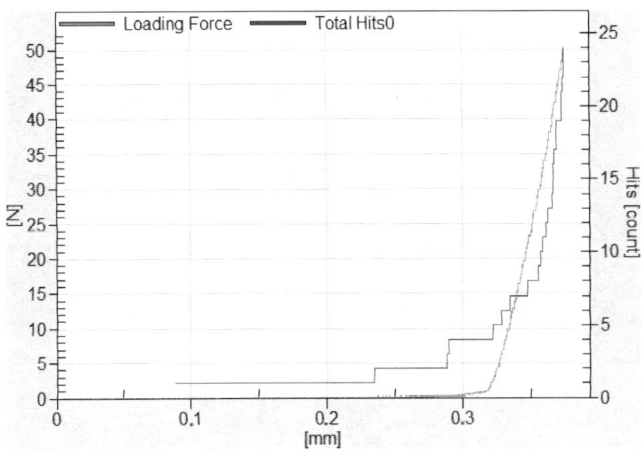

Figure 3. Total Number of AE Hits, Indentation test at F_{max} = 50 N on sample nACRo with BIAS 85V

In Figure 4 can be seen recorded curve of Hits0 signals which was generated from the beginning of indentation up to maximum force 50N. To every acoustic event has been assigned loading force by which it was recorded. The first accorded Hit correspond to indenter bump into surface. From this point

IOP Conf. Series: Materials Science and Engineering **726** (2020) 012004 doi:10.1088/1757-899X/726/1/012004

it is possible observe very small increasing of loading force. Another two Hits can be seen by Force $F_1 = 0.5\,N$. Next acoustic events the system of AE detected at Forces $F_2 = 1.5\,N$, $F_3 = 7\,N$ and $F_4 = 11.5N$. Than start a series of Hits which continuously increasing until maximum load 50 N. Because of very high hardness and very good tribological behaviour of tested coating there are just 23 recorded Hits in total. This is an excellent result and it has been confirmed by AE that this coating is not susceptible for cracking at low loads.

However, even if we use Hit parameter curve it`s not possible exactly determine the critical force of the first crack from Hit curve. There are more hits by different loading forces seen in figure 4. Because the acoustic events could be caused by different physical phenomena which happens into the materials under loading it`s not possible to identify exactly by which force was the first crack generated. From the experience we know that creation and propagation of crack cause relatively big dissipation of energy into material. When dissipated energy from crack creation is recorded then also the amplitude of recorded signal is high and of course the recorded acoustic signal energy must be high as well. In this case another parameter which is evaluated from recorded AE signals can help us to identify the pop-in that corresponds to the first crack. This parameter which can be used for quantification of events registered by system of AE is the Hit Energy and the curve recorded during indentation is shown in figure 5. Using this parameter, it is possible to determine how much energy has been dissipated into the material by crack initiation and growing.

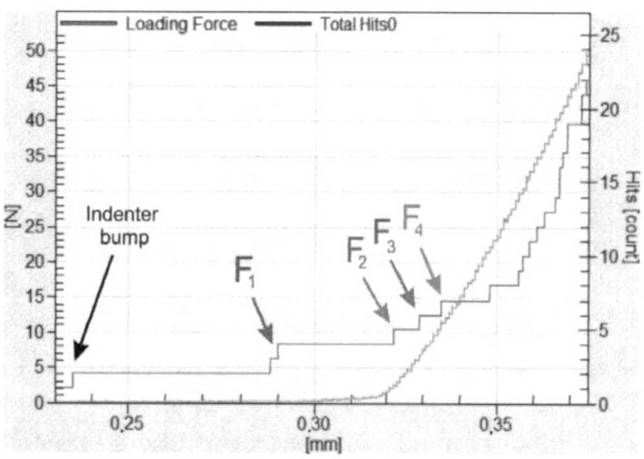

Figure 4. Indentation test at $F_{max} = 50\,N$ on sample nACRo with BIAS 85V - Total number of Hits - zoom.

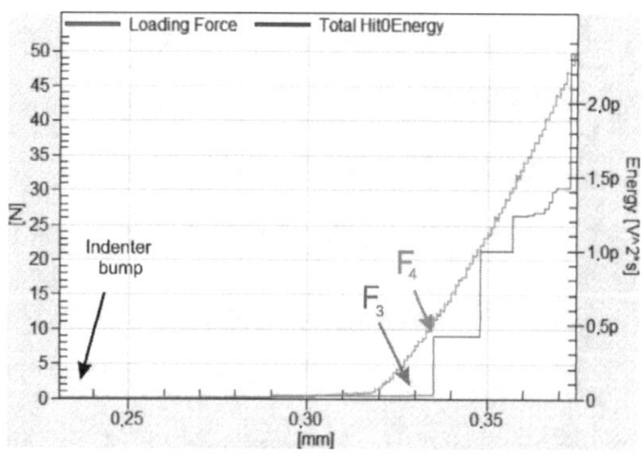

Figure 5. Indentation test at $F_{max} = 50\,N$ on sample nACRo with BIAS 85V - Hits Total Energy - zoom

Figure 6. Indentation test at $F_{max} = 50\,N$ on sample nACRo with BIAS 85V - Picture of imprint scanned by SEM

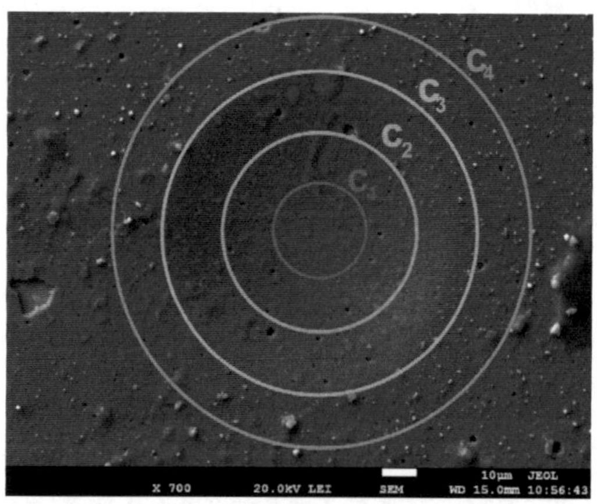

Figure 7. Indentation test at $F_{max} = 50\,N$ on sample nACRo with BIAS 85V - Visualisation of calculated substrate plastic zones c

Table 2. Calculated plastic zones c with different F_C

Max. force (N)	Value of critical Force (N)	Yield stress of ductile substrate σ_y (GPa)	Plastic zone diameter (μm)
$\boldsymbol{F_1}$	0.5	35	26
$\boldsymbol{F_2}$	2.5	35	58
$\boldsymbol{F_3}$	7.0	35	98
$\boldsymbol{F_4}$	11.5	35	125

In the table 2 the plastic zones are calculated according to Equation 1. From the load-displacement curve and related acoustic signal parameters is possible assume that first recorded Hit is not related to creation of crack. The first crack occurs according to Hit Energy curve in the figure 5 by the loading force $F_3 = 7\ N$ shown in the figures 4 and 5.

After the indentation the indenter imprint was scanned by SEM. The picture from SEM of the imprint is in the figure 6. In figure 7, the calculated dimensions of substrate plastic zones c are illustrated. From the figures 6 and 7, it is possible to confirm that around calculated diameter of plastic zone c_3 are located visible coating circumferential cracks.

5. Conclusion

In this study, the sample with nACRo coating deposited on high-speed steel (STN19 852, HSS 6-5-2-5) substrate was investigated during indentation by measuring of AE signals. According to the acoustic emission signals represented by number of hits and the energy of acoustic signals (cracks) recorded during the modified Daimler-Benz adhesion test, we can track creation of cracks in the coating. From the AE data analysis is also possible define when exactly the first crack occurs. Subsequently it is possible to assign what is the value of Force when the first crack occurs. This value is the critical force which is important for calculation of tribological and adhesion parameters of coatings.

The acoustic emission signal detection and evaluation makes it possible to evaluate:

- susceptibility of coating to untimely crack at low loading forces,
- determining of critical force when the first crack occurs.

The aim of this paper is to find a more effective and accurate methodology to determine tribological properties of hard coatings deposited on hard substrates. One of the most outstanding advantages of this methodology is very fast and precise detection of the coating's cracking.

Acknowledgements

The research was supported by the Slovak Research and Development Agency under the contract No. APVV-15-0168, and the Slovak Republic Scientific Grant Agency (VEGA) within grant No. 1/0540/19.

References

[1] *Verein Deutscher Ingenieure Normen, VDI 3198* 1991VDI-Verlag, Dusseldorf

[2] *The VDI 3198 indentation test evaluation of a reliable qualitative control for layered compounds*, https://doi.org/10.1016/S0924-0136(03)00300-5

[3] Fu K, Tang Y, Chang L 2016 *Toughness Assessment And Fracture Mechanism Of Brittle Thin Films Under Nano-indentation* Fracture Mechanics - Properties, Patterns and Behaviours, *Ed.* Lucas Alves, IntechOpen doi: 10.5772/64117

[4] Mercier D 2018 *Pop-In Documentation*

[5] Navamathavan R et al. 2006 *Mater. Chem. Phys.* **99** 410–413

[6] Krabbe M 2014 *Fracture Toughness Of Thin Films Estimated By Rockwell C Indentation* Phd Disertation Thesis, Aarhus University, Department of Engineering, Denmark, ISBN 978-87-93237-27-8

[7] Bahr D F, Woodcock C L, Pang M, Weaver K D, Moody N R 2003 Indentation induced film fracture in hard film – soft substrate systems *Int. J. Fract.* **119/120** 339–349
[8] Malzbender J, Den Toonder J M J, Balkenende A R, De With G 2002 Measuring mechanical properties of coatings *Mater. Sci. Eng.* **R36** 47-103

Influence of focal length on depth of engraved PMMA surface

J Knedlová, L Sýkorová, V Pata and A Škrobák

Tomas Bata University in Zlín, Faculty of Technology, Vavreckova 275, 760 01 Zlín, Czech Republic

knedlova@utb.cz

Abstract. The paper focuses on laser cutting of polymer material PMMA. Specifically, the influence of optics on the engraved surface is investigated with use focal length lenses of 1.5"; 2.5"; 4". Commercial CO_2 laser ILS 3NM by Laser Tools & Technics Corp. Taiwan, was used for experimental machining.

1. Introduction

The selection of optimal technological conditions is dependent on the chosen material. Polymers are one of the youngest groups of construction materials. Chemical substances are those, which have very different properties, and their huge molecules contain mostly the atom of carbon, oxygen, nitrogen and hydrogen. As a final product, polymeric materials are solid, but during processing at high temperature and pressure, they are liquid to assume the future shape of the product. It is possible to modify the properties using various modification and additives. Generally, the plastics are light weighted; they have thermal and electrical insulation properties and no corrosion [1-2].

A polymer is a macromolecular substance formed by the chains of the macromolecules (Figure 1). These are formed by linking the building units by primary covalent bonds and contain a large number of the atoms. Low-molecular substances, so-called mers, which repeated in the chain, are the basic building part. The basic part of the chain forms a carbon atom, which is able to bind to each other to form long chains [1-2].

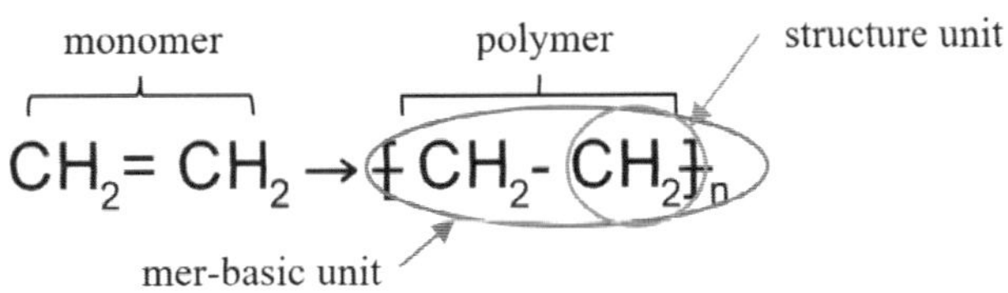

Figure 1. Polymer structure.

The plastics are polymers, which are the in at normal conditions mostly hard. With increasing temperature they become moldable and plastic (hence the name plastics). The plastics are divided into thermoplastics and thermosets. The thermoplastic is able to be repeatable heated and convert to the state of melt or to the viscous flow. By cooling below the melting temperature, which characterizes the semi-crystalline polymers, or below the flow point, which characterizes the crystalline polymers, the thermoplastics can be allow to solidify at temperatures that characterize the given type of

IOP Conf. Series: Materials Science and Engineering **726** (2020) 012005 doi:10.1088/1757-899X/726/1/012005

thermoplastic. Thanks to this ability, thermoplastics are recyclable because there is no structural change [2].

They consist of linear long chain macromolecules. These chains are connectedly by intermolecular interactions (van der Waals forces, hydrogen bonds). When the thermoplastic heats up, the intermolecular interactions decrease and the polymer softens. Thermoplastics are process able materials and can be easily cast or moulded [2].

Laser beam machining (LBM) is one of the most widely used thermal energy based non-contact type advance machining process, which can be, applied for almost whole range of materials. Laser beam, is focussed for melting and vaporizing the unwanted material from the parent material. It is suitable for geometrically complex profile cutting and making miniature holes. Among various type of lasers used for machining in industries, CO_2 laser are most established. In recent years, researchers have explored a number of ways to improve the LBM process performance by analysing the different factors that affect the quality characteristics. The experimental and theoretical studies show that process performance can be, improved considerably by proper selection of laser parameters, material parameters and operating parameters [3-9].

2. Sample preparation

Samples were made using device ILS 3NM, CO_2 laser, having a wavelength λ = 10.6 μm and with a maximum power P = 100 W, maximum feed rate v_f = 1524 mm.s^{-1}. For the depth cut measurement, the PMMA samples with the thickness of 5 mm were created at constant power P = 100 W and variable cutting speed v_f, which changed from 10 – 100 % (Table 1, Figure 2) using lenses with the different focal distance f = 1.5", 2.5" and 4".

Table 1. Feed rate.

Sample		1	2	3	4	5	6	7	8	9	10
v_f	%	10	20	30	40	50	60	70	80	90	100
	mm.s^{-1}	152.4	304.8	457.2	609.6	762	914.4	1066.8	1219.2	1371.6	1524

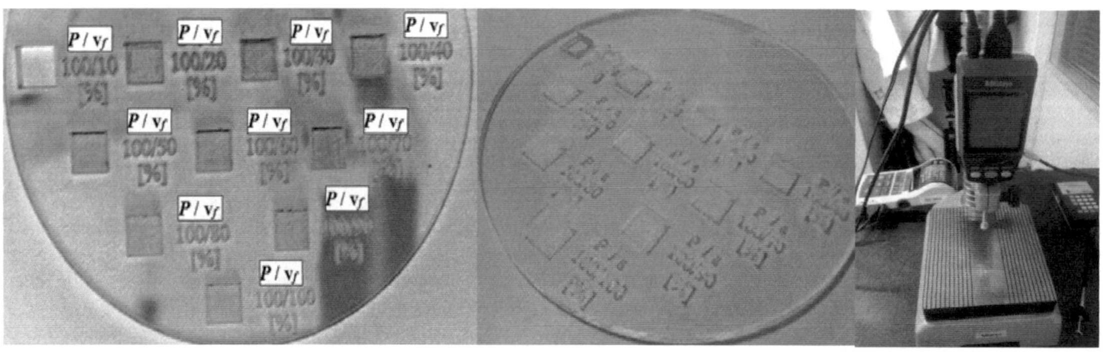

Figure 2. Samples for experiment and measurement device.

3. Depth measurement of cut

The digital gauge H0530 – Mitutoyo with the resolution 0.001 mm and the measurement range from 0 to 30 mm used for the depth cut measurement of the engraving.

In the Table 2, the measured values of depth d of the removed material by ablation are listed in the dependence on the processing conditions, power P and cutting speed for engraving v_f and various focal distance f. The measured values are compared in the followed Figure 3.

Table 2. Summary of results.

Sample		1	2	3	4	5	6	7	8	9	10
v_f	[%]	10	20	30	40	50	60	70	80	90	100
	[mm. s^{-1}]	152.4	304.8	457.2	609.6	762	914.4	1066.8	1219.2	1371.6	1524
f ["]						***d*** [μm]					
1.5		through	2817	1900	1490	1048	1000	889	769	679	619
2.5		through	**2992**	2094	1534	1228	1032	889	759	655	613
4		through	2142	1985	1509	1167	1006	863	738	626	**580**

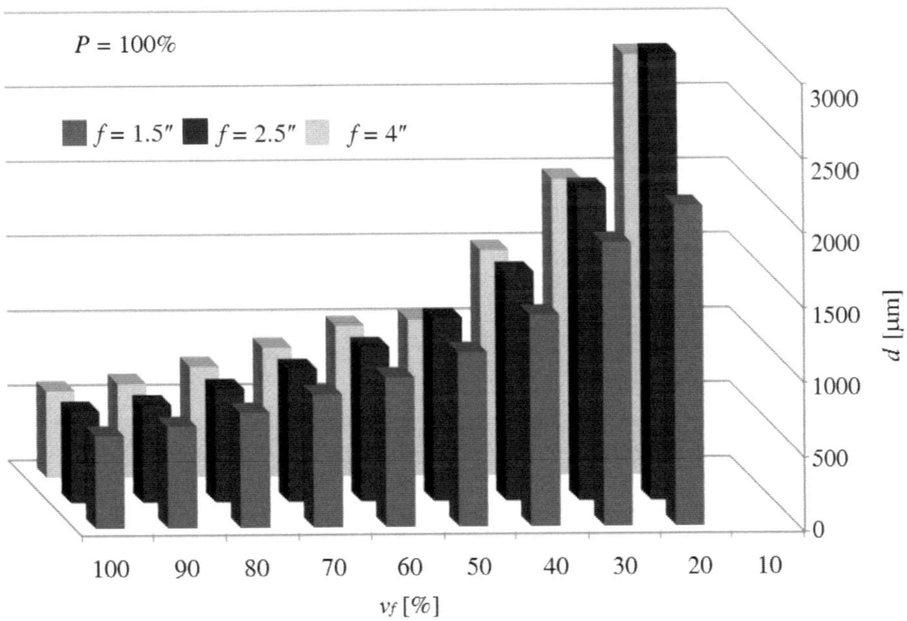

Figure 3. Comparing the depth of cut for different focal lengths on feed rate.

4. Results

From measured results of the removed layer by laser ablation with the depth of d it is possible to see, that during static parameter of power P and variable feed rate v_f, the values changed with the change of focal distance f (Figure 4).

The cutting speed 10 % (15.24 mm.s^{-1}) caused the removal of whole thickness of the sample (5 mm) during the use of all three lenses.

Using the cutting speed from 20 % (304.8 mm.s^{-1}) caused the highest amount of the removed thickness during the use of the lens with the focal distance 2.5".

The lens with the focal distance 4" caused during the engraving from 20 to 60 % of feed rate the increase of the value d compared to the focal distance 1.5". In the range of speeds from 70 % (1066.8 mm.s^{-1}) to 100 % (1524 mm.s^{-1}), the removed material by ablation had the higher values d with the focal distance 1.5" in comparison to 4". At the speed of 1066.8 mm.s^{-1} it was reached the same value of d = 889 μm with 1.5" and 2.5" focal distance.

The greatest depth value d = 2992 μm was reached using feed rate v_f = 304.8 mm.s^{-1} and the lens with the focal distance f = 2.5". The smallest depth d = 580 μm was reached using feed rate v_f = 1524 mm. s^{-1} and the lens with the focal distance f = 4".

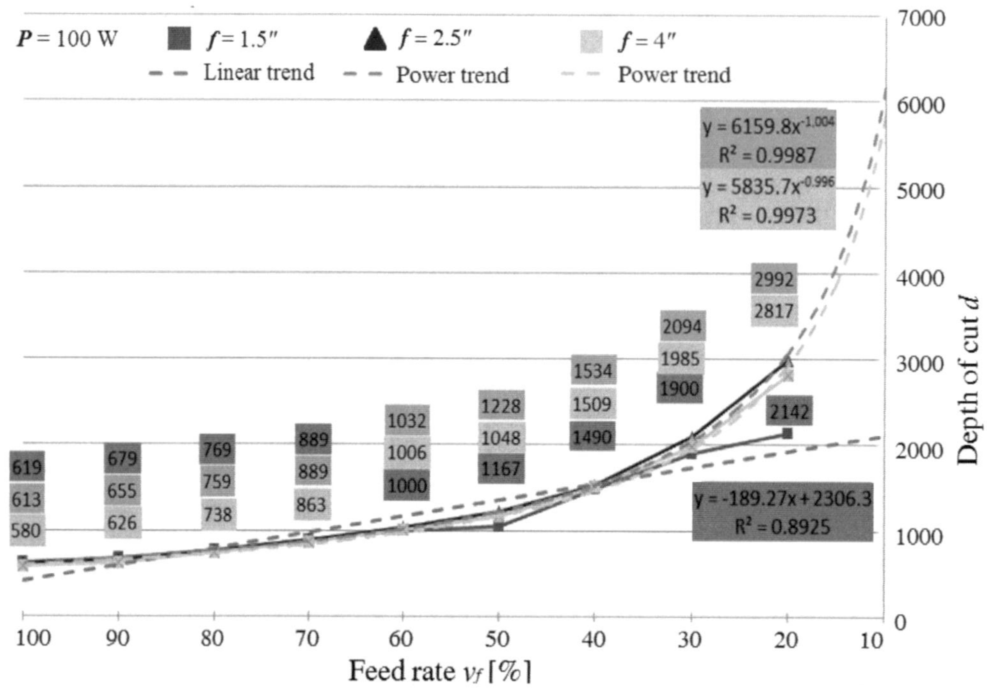

Figure 4. Dependency depth of cut from feet rate.

On the machined surface, at various setting of feed rates and the changing of focal distance, the changes of the sectionalisation surface occur. In Figure 5, the surface with the maximal depth d = 2992 μm and minimal depth d = 580 μm is displayed.

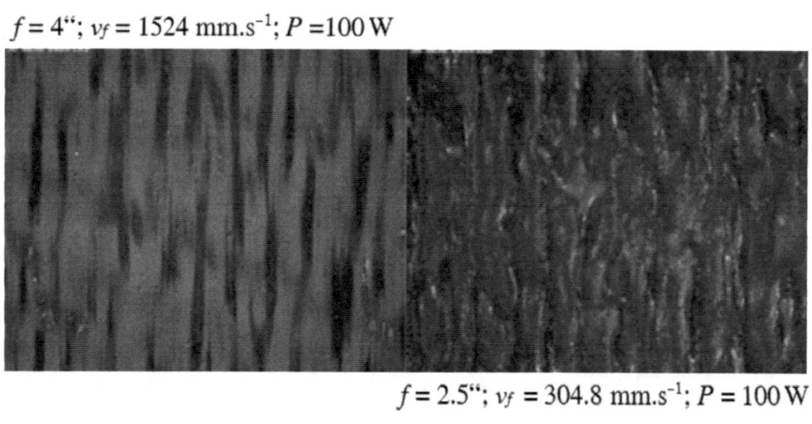

Figure 5. Influence of process conditions on surface structure.

5. Conclusion

The aim of this study was to find out at which cutting condition it is possible to achieve the certain depth using various focal distances. With changed processing conditions, the depth after engraving and the appearance of the surface changed. It is necessary to take in consideration not only the depth in the dependence on the cutting conditions, power P and feed rate v_f, but also the influence of the focal distance f, which, as shown by, causes at PMMA the variable thickness of the removed material in the dependence on the change of the feed rate. All those mentioned factors influence the proper choice of the processing conditions. Trends equations have been establish and in most cases have a power character.

Acknowledgment
This work and the project developed with the financial support of the internal grant of TBU in Zlin No. RVO/FT/2019/003.

References

[1] Ducháček V 2011 *Polymery - výroba, vlastnosti, zpracování, použití* (Prague: Vydavatelství VŠCHT-Czech Republic) ISBN 9788070807880 (in Czech)
[2] Běhálek L 2016 *Polymery* (Prague: Code Creator, s.r.o.) https://publi.cz/books/180/04.html ISBN 978-80-88058-68-7 (in Czech)
[3] Ion J 2005 *Laser Procesing of Engineering Materials – Principles, Procedure and Industrial Application* (Netherlands: Elsevier)
[4] Maňková I 2000 *Progresívné technologie* (Košice: Vienala) ISBN 80-7099-430-4 (in Slovak)
[5] Pata V, Sýkorová L, Knedlová J and Řezníček M 2014 Application of Theory of Hypothesis Testing in Process of Laser Micromachining *Applied Mechanics and Materials* **693** 247-255
[6] *A Deep Dive Into Laser Cutter Speed And Power* 2019 https://makezine.com/2019/03/04/a-deep-dive-into-laser-cutter-speed-and-power/
[7] Hubeatir K A, Al-Kafaji M M and Omran H J 2018 Deep Engraving Proces of PMMA Using CO_2 Laser Complemented By Taguchi Method *IOP Conf. Ser.: Mater. Sci. Eng.* **454** 012068
[8] Kasman S and Saklakoğlu I E 2010 *Investigation of laser deep engraving process: A case study for surface roughness and engraving depth* Int. Conf. on Advances in Materials and Processing Technologies, AMPT2010; Paris; France; 2010 **1315** 1582-1588
[9] Patel S, Patel S B, Patel A B 2015 A Review on Laser Engraving Process *IJSRD – Int. J. for Scientific Research & Development* **3** 247-250

IOP Conf. Series: Materials Science and Engineering **726** (2020) 012006 doi:10.1088/1757-899X/726/1/012006

Study of bending resistance of sandwich structures

O Šuba, M Kubišová, O Šuba Jr., D Měřínská and L Pitnerová

Tomas Bata University in Zlín, Faculty of Technology, Vavrečkova 275, 760 01 Zlín, Czech Republic

mkubisova@utb.cz

Abstract. Article deals with the study of the flexural limit state of sandwich structures based on fiberglass and polymeric foams. Whether geometrical and material parameters influence the resulting load-bearing capacity of these structures are studied experimentally using FEM models. This study shows it is necessary to particular emphasis attention to the issue of flexural strength, the load capacity when of the walls designing sandwich shell products to avoid possible failures in the practical use of these types of structures. Subsequently, Horn's evaluation method is used to compare the experimental testing and the FEM model.

1. Introduction

Production technologies, along with the study of the properties of sandwich structures on a polymeric basis, are steadily rising and are the subject of many researchers. Sandwich structures are generally characterised by excellent mechanical properties at the very low weight. On the other hand, the choice of the sandwich component concept also brings problems associated with its macrostructure [1-3]. As a result of the geometric and material parameters of the sandwich wall, it is layered, and therefore non-homogeneous, element whose some aspects of mechanical behaviour may differ significantly compared to a conventional homogeneous wall. The resulting level of structural strength is determined by factors that generally reduce the theoretical bending strength resulting from the bending theory [4-5]. In particular, a standard three-point test cannot be used to measure the exact bending properties. The reason is mainly the effect of the generally significant shear stiffness, the low values of the shear modulus of the elasticity of the expanded core the resulting useful mechanical behaviour of the sandwich element. In the three-point bending test, in addition to the flexural-normal deformation of the supporting layers, the shear stresses, which are known to be transmitted by virtually only the layer of the expanded core, are involved in the resulting deformations relatively extensively. Due to the low stiffness of the core, the shear deformations are significant, and the test is therefore unusable for determining the bending properties. Also, the results are often devalued by possible indentations at the support sites [6-7].

2. Materials and methods

2.1. Problems of bimodularity of supporting layers

For the pure bending stiffness of the layer-composite cross-section (without the influence of the shear deformations of the core), the modification of the technical bending theory can be laid.

$$K_O = E_+ \, J_R \tag{1}$$

where E_+ is the modulus of elasticity of the surface laminate tensile layer and J_R value represents the quadratic moment of the reduced cross-sectional area of the layer structure to the neutral axis - see Figure 1.

The so-called reduced cross-sectional area of the non-homogeneous-composite element is generally determined by reducing the widths of the individual layers relative to the selected reference (usually the largest) value of the modulus of elasticity of the structure.

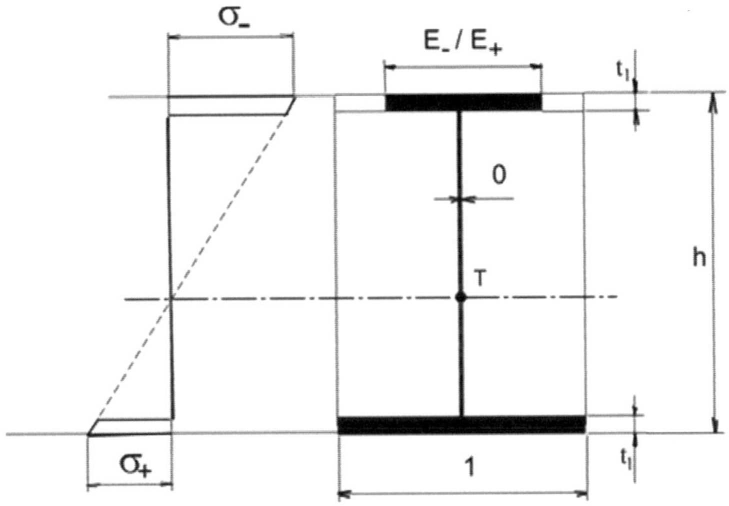

Figure 1. Reduced cross-sectional area and bending stress, considering the bimodularity of the supporting layers.

The existence of carrier bimodularity shows some aspects that negatively affect or reduce the load-bearing capacity of the sandwich structure compared to the expected values according to the conventional technical bending theory. As the floor of Figure 1, due to the asymmetry of the reduced cross-sectional area to the neutral axis, the bending stress in the printed carrier layer is more significant, while the modulus of compression flexibility is less than in the tension. From a mechanical point of view, this is a negative tendency, since it is a mechanical problem of the stability of the printed layer on the elastic core.

2.2. *Numerical analysis of the ultimate state of the flexural load-bearing sandwich structure in bending*

To study the flexural stiffness of the sandwich, wall on the rigidity of the expanded nucleus, linear elastic FEM models of bending tests with a structure corresponding to samples prepared for bending test was used. Solidworks/Cosmos tools software was used to build the FEM model. The model represents a four-point bending test, with the supports and load shown in Figure 2. With this arrangement, the central part is loaded with a constant bending moment and thus with zero transverse forces (pure bend).

The calculation is performed as a modal analysis of a 2D solids. Mechanical data were entered based on experimentally obtained values. In our case, the laminate layer exhibited a tensile modulus of 22 000 MPa and compression modulus of 15 000 MPa.

The solution is performed as a modal analysis of the specially modified four-point bending test model (Figure 2). The detail of generated mesh is shown in Figure 3.The samples are adjusted using additional bars so that a clean bend condition is ensured during the test. There is no rigid bond between the strips and the support layers so that the rigidity of the edge portions is only slightly affected. The model naturally assumes the ideal rigid bond between the layers of the structure, so the result is a theoretical, practically unreachable limit state representing a loss of stability of the laminate layer.

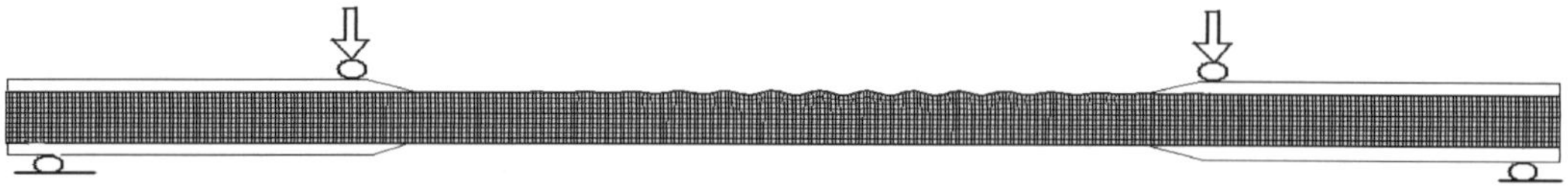

Figure 2. Special arrangement of the four-point bending test.

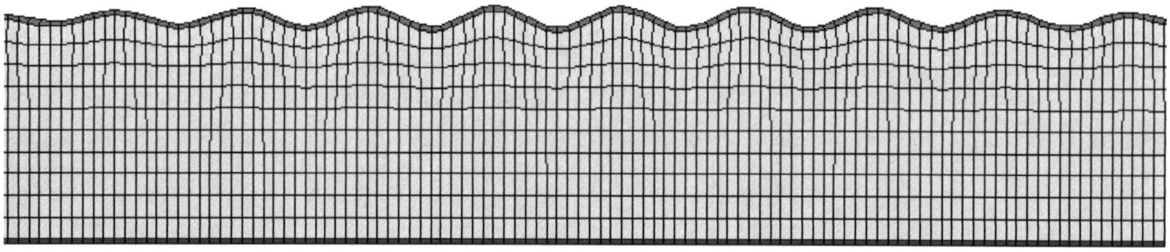

Figure 3. Detail of the model limit state of loss of stability of the printed carrier layer.

2.3. Experimental testing of sandwich structures

When the verification of the results of numerical modelling was prepared, sandwich structure coated glass pre-impregnated fabrics of the basis weight of 296 g.m^{-2}. The impregnating component in the case of this prepreg (PH840-300-42) was a phenolic resin in a total amount of 42 % by weight. Polyethene terephthalate (AIREX T90.100) - core PET foam - a density of 100 kg.m^{-3}, with a modulus of elasticity of 70 MPa, was used as the core of the prepared structures.

From the above materials, sandwich structures were formed, consisting of one, two and three layers of prepreg from the top and bottom and a core of PET with a thickness h = 20 mm. The resin curing technology in the prepreg under the flexible foil in the curing oven was used to produce it. Curing started at 130 °C for 30 minutes, followed by the residence time of 120 minutes at this temperature. After deformation, test bodies were cut from the sandwich panels for standard bending tests.

The results of the load-carrying capacity of the individual structures - with the composition of the bearing layers with the number of mono-layers - lamin 1, 2 and 3 showed a significant dependence on the local value of the adhesion of the layers. The torque values thus varied within a considerable range. Quality values of laminate-foam glued joints were evaluated after each test.

The test consisted of visual inspection of the individual layers, using a microscopic technique and the magnification was optimised 50x [7-8]. It has been found that the consistency of these primarily commercial types of sandwiches showed considerable scattering (from very weak to high-quality cohesion values) and the resulting flexural strength values also corresponded. The theoretically predicted phenomenon of loss of stability was not observed. The results of the measured values of the limit bending moments are plotted for the individual structures (layers with 1, 2 and 3 laminates) in the graph in Figure 4. The results were processed using the Horn method (Table 1).

Table 1. Experimentally determined values of boundary-bending moments of individual structures. The upper value in each column is the theoretical moment determined by FEM by modal analysis.

	Sandwich 1 Slat 1	Sandwich 2 Slats 2	Sandwich 3 Slats 3
Lower limit (Horn minimum) confidential level 5%	24.5 kN	30.7 kN	36.9 kN
Medium (Horn centre)	37.5 kN	53.4 kN	69.2 kN
Upper limit (Horn maximum) confidential level 95%	66.1 kN	69.5 kN	72.9 kN
Theoretical maximum	37.9 kN	71.5 kN	106.5 kN

By mere visual observation, we did not observe the loss of stability due to the corrugation of the printed layer in any specimen, even in samples whose measured, M_{LIM} was relatively close to the theoretical value. We, therefore, used to monitor the moment of the high-speed camera violation, and the tracking results were shocking. The Olympus i-Speed - II high-speed camera was used. Its parameter setting was: frame rate of 10,000 fps (frames per second), resolution of the sensor 800x600 pixels. The image was analysed by internal software, which is part of the supplied high-speed camera system called i-Speed. With samples with excellent consistency, we managed to capture the moment of loss of stability of the printed layer. This effect took place only a very short moment, which the human eye cannot capture. Figure 5 shows behaviour of the sample 0.04 seconds ($t_{(s)}$ = loss of stability – 0.04 sec) before ripple, in Figure 6, then the very phenomenon of loss of stability. Then the destruction of the slate according to occurs (Figure 7).

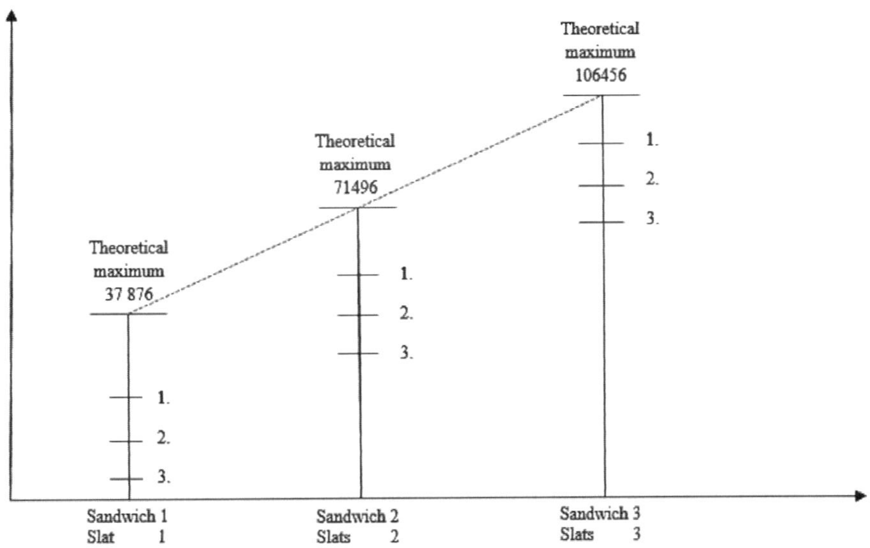

Figure 4. Analysis of small selections.

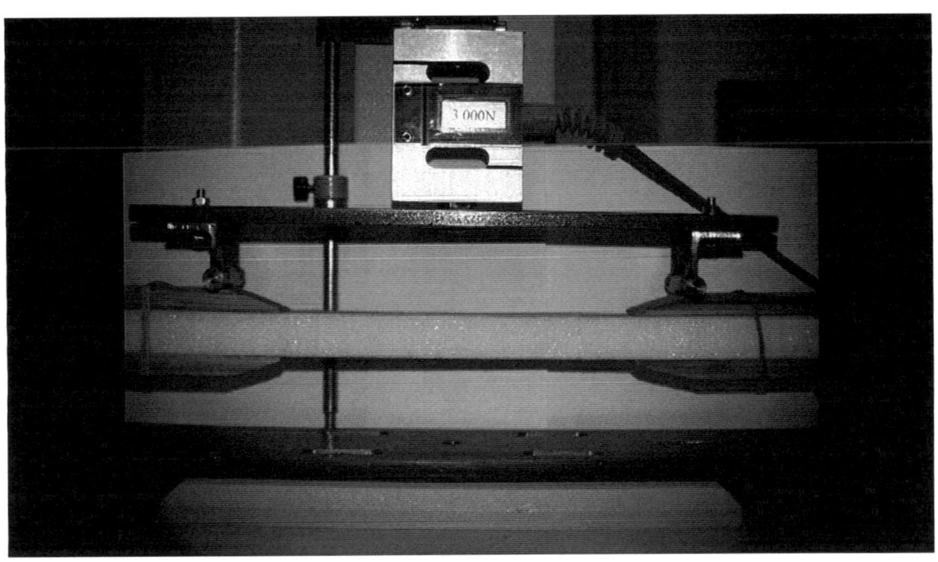

Figure 5. Behaviour of the sample 0.04 seconds before corrugation.

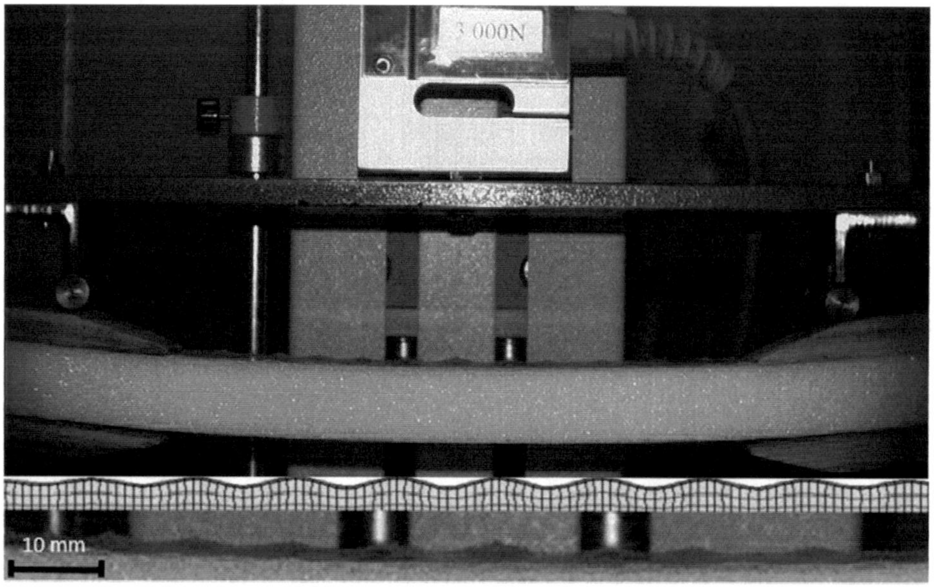

Figure 6. Maximum instability of the sample.

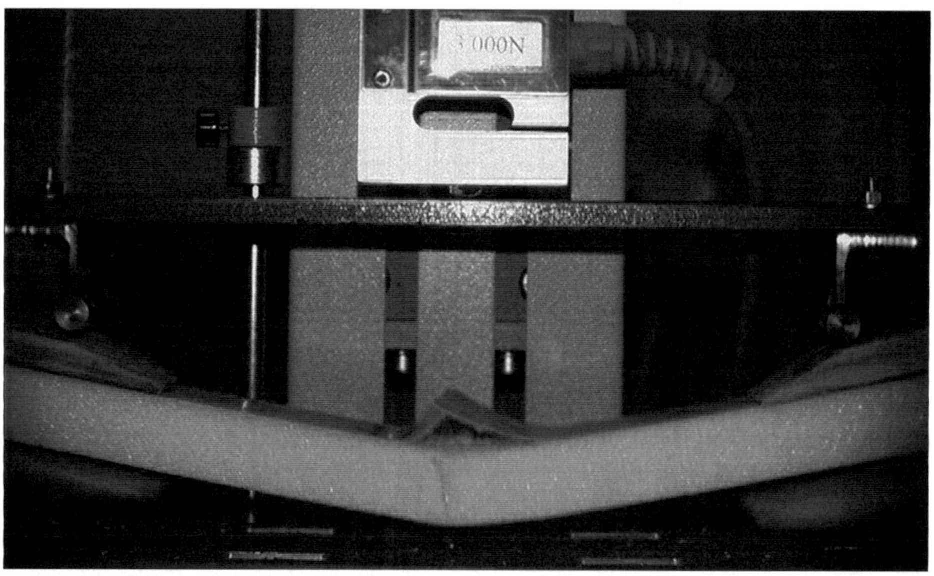

Figure 7. Destruction of the structure.

3. Conclusion

This study showed that particular attention should be paid to the flexural strength issue when designing sandwich products in order to avoid possible failures in practical applications of these types of structures. The limit state mechanism in the bending of sandwich structures based on laminate carrier layers and foam core has been found. It has been shown that the failure state of these sandwich structures represents a short time, in the order of hundredths of a second, in which the first compressed support layer loses its stability and then breaks down. Due to the bimodularity of laminate layers, the load-bearing capacity of sandwiches is always lower due to the lower modulus of the laminate in compression than in tension. Higher modulus values of the tensile layer are never used.

This was followed by verification of the model's self-results, which were achieved using the 4-point bend technique, the respective and described sandwich structures.

The results obtained by the four-point bend were then evaluated by the Horn method, which is recommended for the purpose and frequency of the samples examined [8].

The graphical comparison of the model results and the results obtained with the 4-point bend test was captured using the high-speed OLYMPUS I speed 2 camera system, with a recording speed of 10,000 fps (frames per second). At the point in time $t = 0.00$ s, the most significant match between the model and the real record occurred as the optically based image.

Acknowledgement
This work was supported by an internal grant of UTB in Zlín RVO/FT/2019/003 financed from Funds of specific academic research.

References

[1] Zenkert D 1997 *The Handbook of Sandwich Construction* (Worcestershire: EMAS Publishing) ISBN 978-094-7817-961
[2] ASTM C393/C393M-11 2012 *Standard Test Method for Core Shear Properties of Sandwich Constructions by Beam Flexure*, ASTM International.
[3] PH840-300-42 2014 *Material datasheet* Gurit. [online]. © 2014 [cit. 2018-10-18]. *Available from:* http://gurit.fangle.co.uk/files/documents/PH840-300-42.pdf
[4] Vinson J R and Sierakowski R L 2004 *The Behavior of Sandwich Structures of Isotropic and Composite Materials* (New York: Kluwer Academic Publishers) ISBN 1-4020-0904-6
[5] Gibson L J and Ashby M F 1999 *Cellular solids: structure and properties* (Cambridge: Cambridge University Press) ISBN 0521499119
[6] Valasek P and Muller M 2012 Polymeric particle composites with filler saturated matrix. *Manuf. Technolog.* **12** (13) 272-6
[7] Banghai J, Zhibin L and Fangyun L 2015 Failure mechanism of sandwich beams subjected to three-point bending *Compos. Struct.* **133** 739-45
[8] Meloun M, MILITKÝ M A and Forina M 1992 *Chemometrics for analytical chemistry* (New York: Ellis Horwood) ISBN 01-312-6376-5.

IOP Conf. Series: Materials Science and Engineering **726** (2020) 012007 doi:10.1088/1757-899X/726/1/012007

Analysis of behavior of composite material during loading tests

V Mára[1], J Krčil[1] and M Černý[2]

[1]Czech Technical University in Prague, Faculty of Mechanical Engineering, Department of Materials Engineering, Karlovo náměstí 13, 121 35 Prague 2, Czech Republic
[2]Czech Technical University in Prague, Klokner Institute, Solinova 7, 166 08 Prague 6, Czech Republic

vladimir.mara@fs.cvut.cz

Abstract. Composites are widely used as a material suitable for construction applications, but the operating conditions can lead to reduction of materials properties, damage initiation and collapse of the structure, so it is necessary to monitor condition of components and thus prevent its catastrophic failure. The unidirectional glass fiber reinforced polymer matrix composite (GFRP) was inspected with usage of non-destructive (NDT) acoustic emission during the static loading tests. The 0° test specimens were manufactured and tested under tensile and three-point flexural load for which the custom-made apparatus was used. Failure mechanisms were inspected by detailed analysis using scanning electron microscopy (SEM) to deduce the damage sequence. Tensile stress-initiated matrix cracking which induces formation of scarps, ribbons and riverlines followed by delamination and fiber failure with formation of radials. Flexural load showed cracks in matrix, delamination and typical compression regions in form of fiber microbuckling. Based on the speed of wave propagation and linear localization, damage processes were detected in form of lateral cracking, delamination and integrity loss during the tensile and flexural tests. The data obtained from mechanical testing were correlated with selected acoustic emission parameters (counts, events, amplitude and duration). Behavior of GFRP material during the tensile test can be described very well with usage of counts while the flexural test provides much less information. The evaluation of damage processes using event parameter based on the amplitude and duration proved to be problematic due to the noise interference. The Short-time Fourier transform (ST-FFT) and peak frequency was used for identification of failure modes. In case of tensile and flexural load the matrix cracking has lower frequencies while delamination and fiber failure have higher values.

1. Introduction

Nowadays the application of fiber reinforced polymer composite materials (FRP) is widespread in various industry as suitable material for construction. Due to good mechanical properties (strength, stiffness), low density and corrosion resistance, its usage is mainly in aerospace, aviation, automotive and military [1, 2]. Due to complex nature of composite material, it is difficult to detect damage modes inside the material. The resolution power of many common NDT technique used for monitoring of damage progress is limited, because they can only detect the defect of a certain dimension [3]. The acoustic emission method does not have this limitation and thus is more suitable for damage analysis than other techniques.

1.1. Acoustic emission

The acoustic emission (AE) is NDT method for materials testing which have great impact in the area of real-time in-situ health structure monitoring. This technique measures and analyzes elastic stress waves generated by formation or propagation of defect. Ability to detect and localize defect inside the material helps to secure the safety of construction parts during operation and extends their lifetime, since many defects are irreversible [4]. Most of the defects are of outer or inner character (on the surface or inside the material) or they might be present at difficult-to-access places so it is often difficult or impossible to use other common NDT methods like ultrasound or X-ray [5]. The common approach for evaluation of damage processes inside the polymer composite material is usage of acoustic emission parameters, such as events, counts or energy [6–8]. As many authors have stated, the increase of cumulative values or change in their activity is tied to damage mechanisms initiated in material due to the applied stress. For example, at the beginning of a load, during the matrix cracking, the overall increase of signal activity is small because less energy is needed for crack growth and propagation. The amplitude is usually low as well while fiber related damage processes have higher amplitude and signal activity, since a larger amount of energy is needed [9]. Popular approach is waveform analysis [10] where during the matrix cracking the amplitude is lower and duration time is also shorter, while for fiber damage the amplitude and duration time is higher [4]. The peak frequency of an AE signal is another popular technique for signal classification. It defines specific range of frequencies for each of damage modes. [7,11,12] The efficiency of acoustic emission is based on sensitivity to numerical methods used for evaluation of measured data. Many of transition AE signals that exists inside the materials structure are not tied to real defects but to other types of processes i.e. background noise. It is necessary to determine the characteristics which will differ the sought phenomenon from undesirable one. For example, crack is indicated by medium or higher value of amplitude with higher duration (based on type of material) while burst signal with lower than three counts and low duration is undesirable noise. [9,13]

The aim of this paper is to perform fast AE damage analysis without using the advanced techniques with correlation of events based on the amplitude distribution and duration, which is supported for comparison by standard evaluation techniques.

2. Methods

2.1. Material

The unidirectional (UD) reinforced epoxy composite was used for tensile and three-point flexural tests. Unsaturated polyester resin was reinforced with E-glass fibers with diameter range of 12-20 μm. The test specimens were cut from 1.5×1.5 m composite plate with variable thickness of 6 mm. The plate was manufactured by hand lay-up method. The vacuum was applied to spread the resin throughout the fiber. The curing process of resin was done at ambient temperature. The quality control and determination of volume fiber fraction was done according to standard ČSN 64 4002 and then compared with image analysis made by the analytic software NIS Elements AR and light microscope Carl Zeiss Neophot 32. The volume fraction of fibers was determined to be 45 ±3% and matrix porosity in form of voids and bubbles is 3.42 ±0.5%. During the image analysis (Figure 1), the irregular distribution of fibers in matrix was found in form of resin-rich areas which may potentially act as a site for damage initiation.

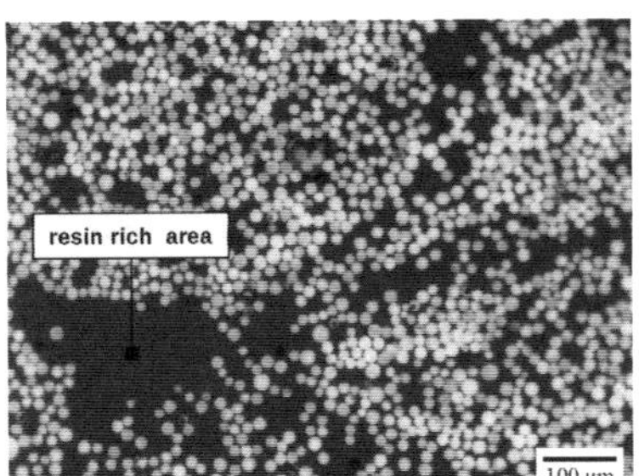

Figure 1. GFRP microstructure (etched by hydrofluoric acid).

2.2. Mechanical testing

The unidirectional reinforced test specimens with 0° fiber (fibers are parallel to the tensile load) orientation were manufactured in accordance with ASTM D3039 [14] and ASTM 7264 [15] for each type of test with dimension of tensile test specimens 200 × 20 × 6.5 mm and 85 × 20 × 6 mm for flexural testing. Tabbing material was not applied on the tensile specimen ends because it was not possible to ensure the efficient adhesive and the tabs tended to debond during the tensile test. The thickness of the specimen could not be changed, so the specimens remained without the tabs. Tensile test was carried out on 200 kN static and 100 kN dynamic universal testing machine Instron 1273 with crosshead displacement rate set to 0.5 mm·min^{-1} at temperature 19.2 °C and 30% R.H. For three-point flexural test, the custom-made apparatus was used (see Figure 2), where movement of transducer ABM U9B is realized by rotation of M5 thread. M5 is set to motion with pair of meshing gears where the pinion is mounted on the shafts output. By stepper engine rotation, the gear and pinion also rotate and simultaneously sets in motion the M5 thread with transducer ABM U9B which acts as a loading element and simultaneously as a measuring load force. Deflection is measured by LVDT E-300 transducer mounted opposite to ABM transducer. To control the apparatus the Data Acquisition/Control Unit HP 3852A was used together with set of digits integrating voltmeter HP 44701A, analogue output arbitrary waveform DAC HP 44726A and HP Agilent 44709A. In the development environment LabWindows™/CVI, the software was written for data acquisition and apparatus control. The displacement rate was set to 0.7 mm·min^{-1} (equal to 1 rpm) and span to 63 mm. The test was performed at temperature 22.7 °C and 36% R.H. (relative humidity).

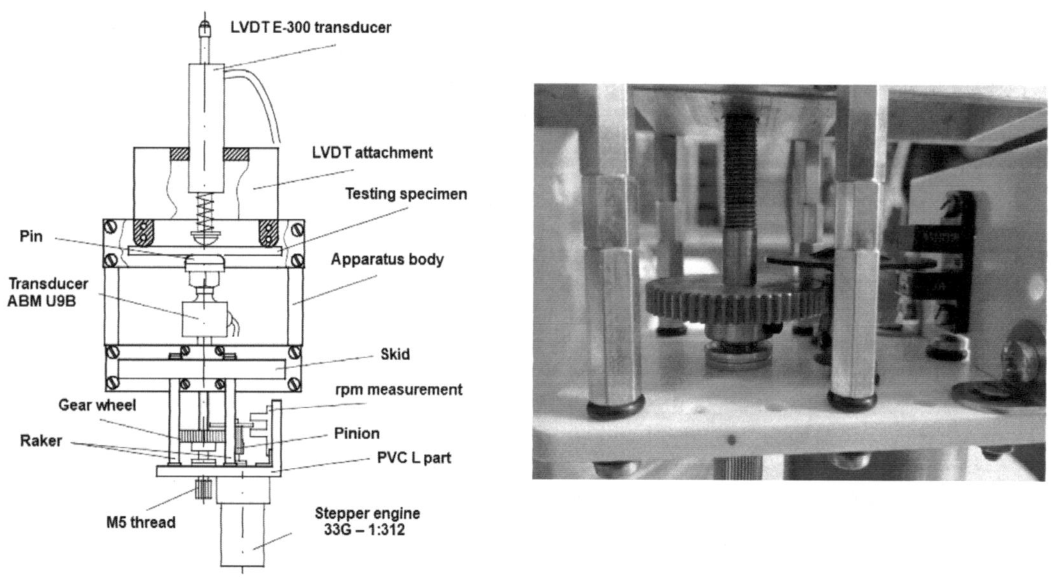

Figure 2. The apparatus scheme for three-point flexural test together with control detail.

2.3. Acoustic Emission

The acoustic signal was being monitored during the mechanical tests by PCI-2 based AE system from Physical Acoustic Corporation (PAC) with pair of differential wideband sensors (100-900 Hz). Each sensor was connected to EPA 1220A preamplifiers with 40 dB gain and mounted on the surface of testing specimens (see Figure 3) with M Bond 200 adhesive kit in the distance of 60 mm for tensile specimen and 40 mm for flexural specimen. For a noise removal the signal passed 20 kHz low-pass filter and 1000 kHz high-pass filter. The threshold type was fixed with value of 30 dB and sample rate was set to 4 MSPS. The AE signal was analyzed by using the following parameters: cumulative counts, amplitude, events and duration time. Analyzing the waveform, the frequency spectra was obtained.

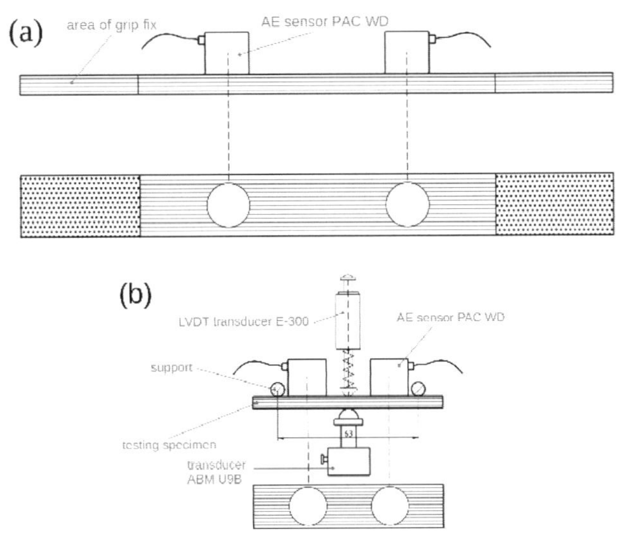

Figure 3. The arrangement of AE sensors: (a) Tensile test, (b) Flexural test.

For the linear localization of AE signal, the velocity of waves spreading through the testing specimens was measured by Krautkamer USM 25 Ultrasonic Flaw Detector equipped with dual 2 MHz SEB-2 KF5 probe. Results were compared to the Hsu-Nielsen pencil lead break test which was performed by breaking the 2H Ø0.5 mm lead on the surface of testing specimens to detect acoustic emission signal. Based on the difference in time of arrival, the wave velocity was calculated. The elastic wave propagates in GFRP material with velocity of 4 500 $m \cdot s^{-1}$ for both types of testing specimens. The data correlation with tensile test was used to identify the failure sources.

2.4. Fractography
The failure mode after tensile and flexural test was evaluated according to ASTM standards and described as three-part failure code. The fracture surfaces were sputtered by chromium evaporation using vacuum evaporator JEOL JEE-4X with current set to 50 A. Scanning electron microscope (SEM) JEOL JSM-7600F has been used for damage process evaluation. Ultra-low voltage gentle beam (GB-low) method with acceleration voltage set to 1 kV and secondary electrons (SEI) were used for fracture surface observation.

3. Results and discussion
3.1. Mechanical testing
Two main failure mechanisms can be observed by naked eye after tensile test. The first are lateral cracks on the surface layer (gelcoat) accompanied by smaller capillary lateral cracks (see Figure 4a). Cracks are under approx. 45° which corresponds to shear stress component, where the highest stress value is under 45°. Lateral cracks are followed by the delamination which initiates in the area of AE sensor and propagates through the grip area (see Figure 4b) towards the specimen edge. The final code was determined as M(ALG)MV where multiple failure (M) type was found in form of angled (A), lateral (L) and grip (G). Failure area was also evaluated as multiple (M) with various (V) location.

For three-point flexural test the failure occurred under the loading pin. Figure 4b shows that lateral cracks appeared on the surface layer before the main failure initiation. In this case, the code is M(TB)AV, where the failure mode is multi-mode (M) with tension (T) and compression (C) with failure area at loading nose (A) and various failure location (V). The ultimate tensile strength is 390 MPa, while flexural strength is 224 MPa.

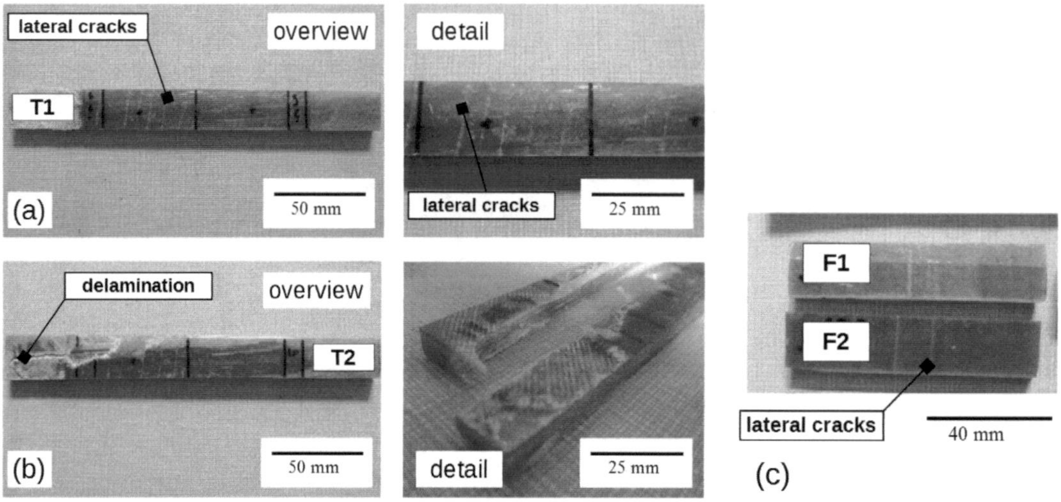

Figure 4. The damage evaluation after mechanical tests: (a) Lateral cracks on the surface, (b) Delamination along the fibers, (c) Lateral crack and failure in the middle of test specimen.

3.2. Fractographic analysis

Both fiber and matrix dominant fractographic features were observed on fracture surfaces and some of them are similar for both tensile (Figure 5c-f) and flexural load (Figure 5a-b), e.g. fiber fracture or matrix microflow. The axial tension led to initiation of intralaminar failure mechanisms in form of matrix cracking in the resin-rich areas. The interaction of multiple cracks leads to microflow of matrix and formation of fractographic features, such as scarp, ribbon and riverlines.

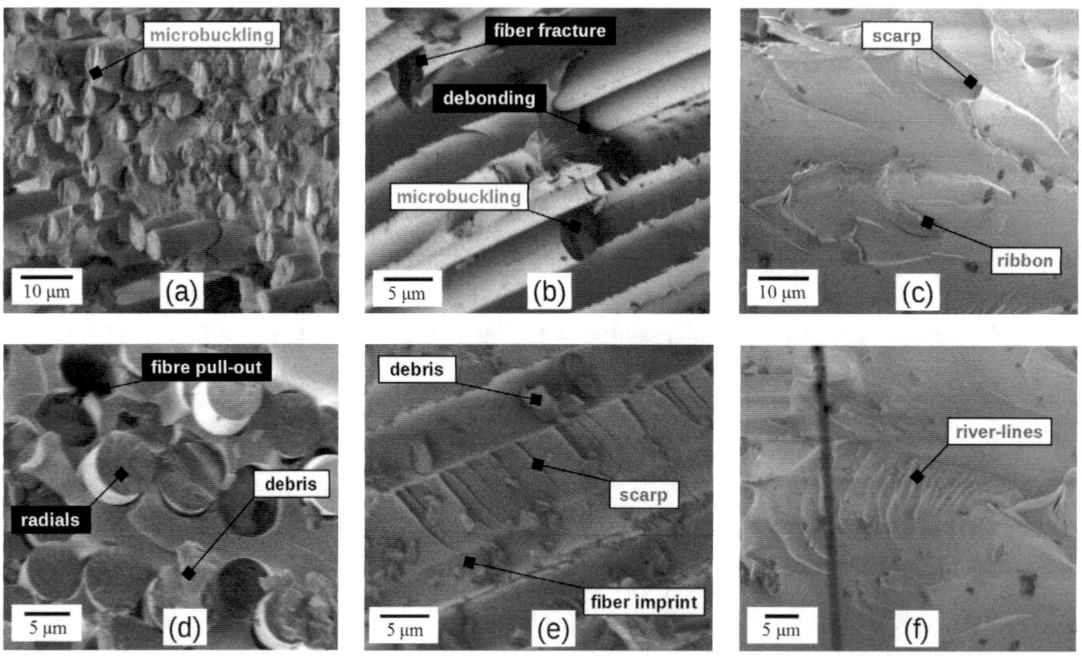

Figure 5. SEM fractograph of flexural and tensile test failure: (a) Microbuckling due to shear stress, (b) Fiber failure, (c) Matrix microflow, (d) Brittle fracture and fiber related failure, (e) Matrix failure mechanism at fiber imprint area, (f) Matrix failure mechanism at resin rich area.

Scarp formation (Figure 5c) not only shows that matrix has less stiffness, but their interconnection creates riverlines (Figure 5f) that can be used for determination of crack propagation. Scarps were also detected next to fiber imprints (Figure 5e) which indicates that crack reached the matrix/fiber interface followed by initiation of shear stress at the interface, this is confirmed by presence of riverlines. The result is either separation of fiber from matrix/debonding (Figure 5b), fiber pull-out (Figure 5d) or brittle fracture and formation of radials (Figure 5d). The fiber pull-outs with fiber fracture were observed infrequently, while the formation of radials is dominant morphological character, which corresponds to failure of 0° fibers. The presence of high amount of fiber imprint and scarps was also observed (Figure 5e) in the case of delamination failure.

Both tension a compression regions can be observed on the fiber fracture surface in case of flexural load. It is a result of fiber buckling due to induced compression stress where the side support of matrix is depleted, and fiber failed by microbuckling (Figure 5a-b). Since the matrix has low resistance in compression, the cracks initiate first then followed by the fiber dominant failure in a form of fiber fracture and microbuckling. Matrix has brittle behavior in flexural load as well which is confirmed by presence of riverlines (Figure 5f).

3.3. Acoustic emission

3.3.1. Linear localization. Figure 6 illustrates the results of linear localization during mechanical tests. This technique usually only determines the 2D coordinates of acoustic emission signal source, but certain patterns can be deduced. During the tensile test, the lateral cracks under 45° are induced in the gelcoat layer. These cracks are evenly distributed on the surface and similar distribution can be seen in the graph (Figure 6a). The event rate is also increasing towards the edge, where the sensor is mounted with high peak of event rate around 50 mm. This probably indicates initiation and propagation of delamination which ends with integrity loss and failure of testing specimen. Similar approach can be used in case of flexural load (Figure 6b). Large Gaussian event rate is in the middle of monitored area (around 20 mm) with occasional peaks around. The specimen failure occurred in the middle of testing specimen which corresponds with high rate of events between the sensors.

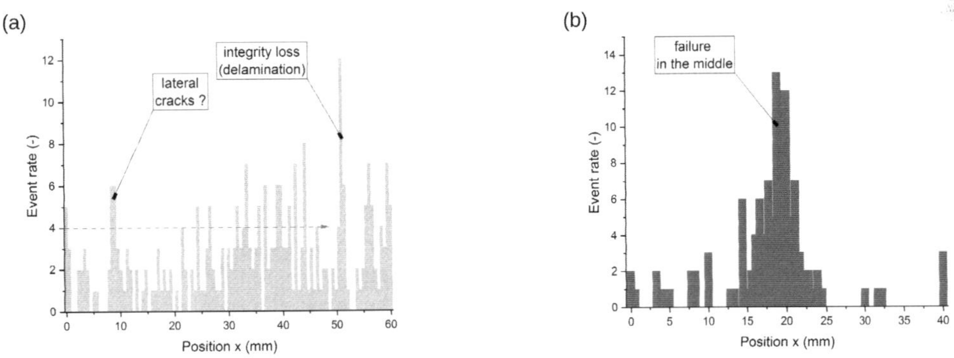

Figure 6. The linear localization of AE source: (a) Tensile test, (b) Flexural test.

3.3.2. Counts. FFigure 7 illustrates the activity of AE signal during the mechanical tests which in case of tensile test creates three separated areas (Figure 7a). The high activity of counts can be seen at the early stage of loading with corresponding sharp increase of cumulative values. The lateral cracks and smaller lateral capillary cracks in surface layer appeared almost at the beginning of test and this type of damage process and can be assigned to the first area up to 7 kN in similar way as stated in [7]. This is followed by decrease of count rate and low signal activity with occasional peak increase and slight increase can be seen in case of cumulative counts and there is no change in profile till reaching the value of 11 kN. This corresponds to matrix cracks, since fractographic analysis showed that matrix cracks are induced at the beginning and need less energy to propagate. Final area begins with large increase in count rate together with sharp increase in cumulative values. Distribution of count rate is similar as it was with events at linear localization. The fibers begin to break or are pulled-out. The slight load decrease at 44 kN indicates initiation of delamination and gradual loss of integrity, which corresponds with high count rate. It is similar to delamination initiation and parameter distribution in [9]. The activity drops together with load at 49 kN where fracture occurred.

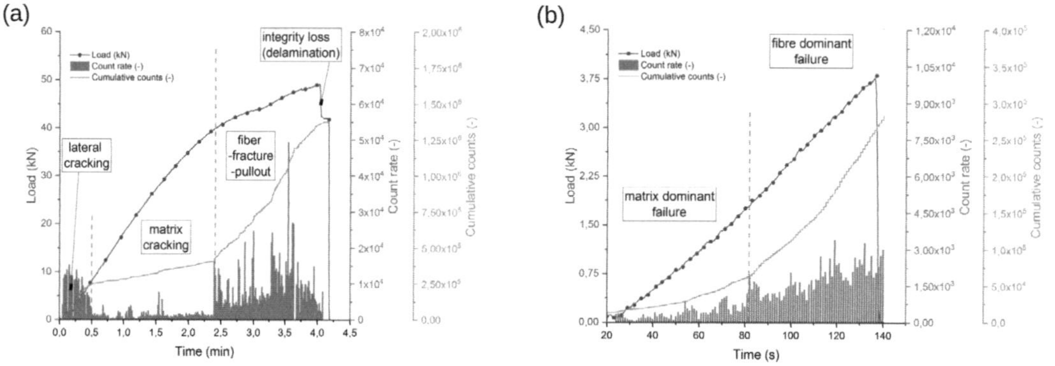

Figure 7. The correlation of load and count parameter: (a) Tensile test, (b) Flexural test.

Completely different behavior can be seen in case of flexural test (Figure 7b) where cumulative counts have almost exponential type of profile. There are still two distinguishable areas. Low activity and slow increase in cumulative counts is sign of failure associated with matrix, since it occurred as first, while the change at 0.7 kN in cumulative counts and almost linear increase in signal activity indicates fiber failure. No decrease in load during the test was observed so it is difficult to distinguish other mechanism types.

3.3.3. Amplitude. Figure 8 shows correlation of acoustic emission events with amplitude and their classification according to the event duration. Based on the claim that shorter duration is associated with matrix damage and higher with fiber dominant, some amplitude bands can be created (see Figure 8a) for a tensile test. The damage associated with fiber can be found between 70-87 dB since the cluster is clearly distinguishable. Matrix damage however might either be in band of 60-70 dB and 87-100 dB (red color) or in 67-75 dB and 84-100 dB (orange). Second variant is more probable since the first one has too low duration and could be exchanged for noise. The amplitude distributions corresponds to Ref. [12]. The evaluation of flexural load is more difficult because there are no separated areas and distribution is completely different (Figure 8b). Red color can be again considered as noise and therefore the matrix failure (orange) is in band of 45-85 dB. It seems that in case of flexure, the damage associated with fiber initiate together with matrix at lower amplitude bands (40-50 dB) and can be also found at higher values (85 dB) which corresponds to amplitude distribution in Ref. [9].

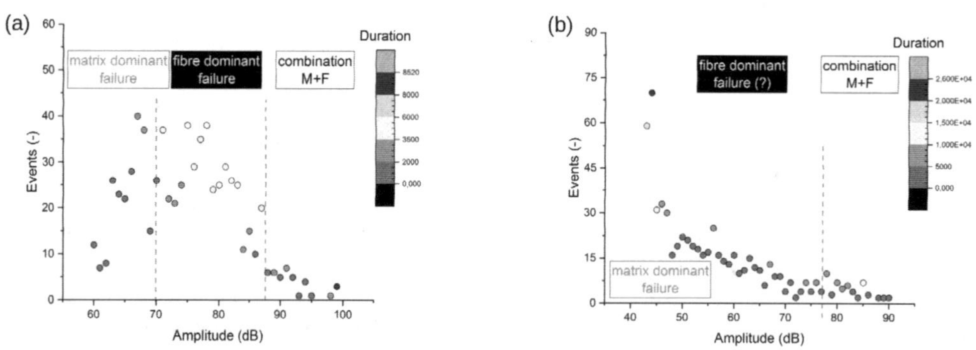

Figure 8. The event rate based on Amplitude and event duration: (a) Tensile test, (b) Flexural test.

3.3.4. Peak frequencies. Decomposing waveforms by fast Fourier transform (FFT) and converting the time to frequency domain (Figure 9) is more effective method to identify the failure sources. The dominant frequencies at 50-150 kHz were observed at lower stresses for tensile test (Figure 9a). Therefore, low frequency events can be attributed to matrix-related mechanisms. Higher frequency belongs to interfacial debonding in form of delamination (200-300 kHz) while fibre damage is at 450 kHz. These frequencies assigned to damage mechanisms are in well accordance with Refs. [3,12,16].

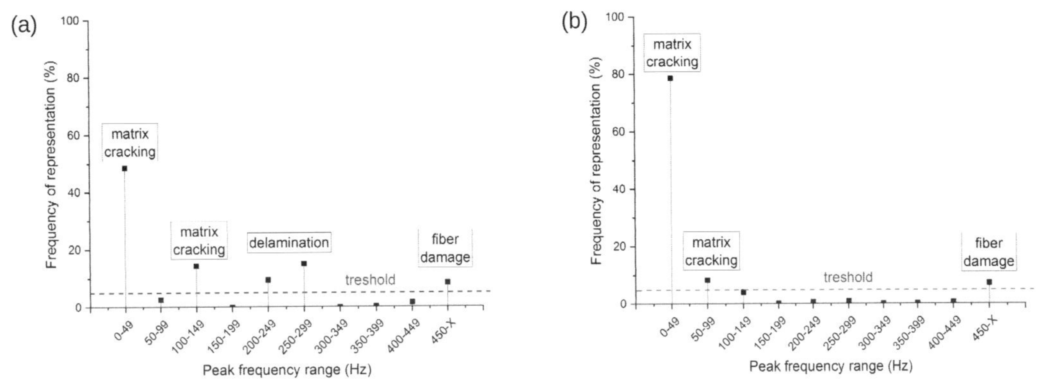

Figure 9. Peak frequencies: (a) Tensile test, (b) Flexural test.

Similar approach can be used for flexural stress (Figure 9b). Highest frequency has matrix cracking at 50-100 kHz while fiber damage is at 450 kHz. No other damage type could be evaluated, since the rest of frequencies was below the threshold (lower than 5 %).

4. Conclusion

The UD testing specimens with 0° fiber orientation were tested under tensile and flexural load. Good results were obtained from flexural test with usage of custom-made apparatus so it can be further used and modified for acoustic emission analysis. Fiber and matrix damage processes were identified by SEM. From the observed fractographic features it was possible to deduce the sequence of damage in the composite microstructure during the loading tests. Linear localization provided not only coordinates for identification of acoustic emission sources, but it was also possible to assign lateral cracks in surface layer to event rate activity for both type of stresses. Based on the events distribution, the delamination at tensile failure could be also identified. Good data correlation was observed for damage mechanisms in case of tensile test. Matrix cracking, delamination and fiber related failure can be detected with counts and evaluated based on the rate and cumulative values. However less information was obtained from the flexural load where only areas related to matrix and fiber damage process were identified and thus no detailed information about specific damage processes could be determined. Amplitude and peak frequency distributions also provided another possibility to identify the failure sources. The evaluation of amplitude bands proved to be problematic. The assignment of matrix cracking to events based on the duration can be exchanged for background noise while fiber dominant failure is easier to identify since event duration is much higher. The distribution is completely different for tensile and flexural stress and it is difficult to determine correct amplitude distribution, so this approach seems to be less reliable. FFT analysis supported the identification of the failure mechanisms with easier distinguishing of damage processes. Matrix cracking has low frequency, while delamination and fiber related damage have higher values. The investigation showed that acoustic emission is suitable method for detection and identification of damage and failure mechanisms in UD GFRP composite, but it is necessary to choose appropriate parameters for data correlation. Higher attention to background noise should be paid since it can be easily confused with event related to damage process.

To improve damage detection by AE the following research will focus on usage of other signal parameters, like absolute energy and RMS (Root Mean Square). In addition to the correlation with load, the parameters will also be correlated with deformation. Further attention will be given to the waveform analysis using Continuous Wawelet Transform (CWT), Fourier Transform (FT) or Discrete Wavelet Transform (DWT).

Acknowledgement
This work was supported by the Grant Agency of the Czech Technical University in Prague, grantNo.SGS17/177/OHK2/3T/12.

5. References

[1] Gay D 2014 *Composite Materials: Design and Applications* (Boca Raton: CRC Press)
[2] Breuer U P 2016 *Commercial Aircraft Composite Technology* (Switzerland: Springer International Publishing)
[3] Ramirez-Jimenez C R, Papadakis N, Reynolds N, Gan T H, Purnell P and Pharaoh M 2004 Identification of failure modes in glass/polypropylene composites by means of the primary frequency content of the acoustic emission event *Compos. Sci. Technol.* **64** 1819–27
[4] Mechraoui S-E, Laksimi A and Benmedakhene S 2012 Reliability of damage mechanism localisation by acoustic emission on glass/epoxy composite material plate *Compos. Struct.* **94** 1483–94
[5] Society for Non-Destructive Inspection 2016 *Practical Acoustic Emission Testing* (Springer Japan)
[6] Njuhovic E, Bräu M, Wolff-Fabris F, Starzynski K and Altstädt V 2015 Identification of failure mechanisms of metallised glass fibre reinforced composites under tensile loading using acoustic emission analysis *Compos. Part B-Eng.* **81** 1–13
[7] Baker C, Morscher G N, Pujar V V and Lemanski J R 2015 Transverse cracking in carbon fiber reinforced polymer composites: Modal acoustic emission and peak frequency analysis *Compos. Sci. Technol.* **116** 26–32
[8] Stankevych O and Skalsky V 2016 Investigation and identification of fracture types of structural materials by means of acoustic emission analysis *Eng. Fract. Mech.* **164** 24–34
[9] Fotouhi M, Suwarta P, Jalalvand M, Czel G and Wisnom M R 2016 Detection of fibre fracture and ply fragmentation in thin-ply UD carbon/glass hybrid laminates using acoustic emission *Compos. Part A Appl. Sci. Manuf.* **86** 66–76
[10] Maillet E and Morscher G N 2015 Waveform-based selection of acoustic emission events generated by damage in composite materials *Mech. Syst. Signal Process.* **52–53** 217–27
[11] Maillet E, Baker C, Morscher G N, Pujar V V and Lemanski J R 2015 Feasibility and limitations of damage identification in composite materials using acoustic emission *Compos. Part A Appl. Sci. Manuf.* **75** 77–83
[12] Chandarana N, Martinez-Sanchez D, Soutis C and Gresil M 2017 Early Damage Detection in Composites by Distributed Strain and Acoustic Event Monitoring *Procedia Eng.* **188** 88–95
[13] Mix P 2005 *Introduction to Nondestructive Testing: A Training Guide* (New Jersey: Wiley)
[14] ASTM 2014 *ASTM D3039 / D3039M-14, Standard Test Method for Tensile Properties of Polymer Matrix Composite Materials* (ASTM International, West Conshohocken, PA)
[15] ASTM 2014 *ASTM D7264/D7264M – 15 - Standard Test Method for Flexural Properties of Polymer Matrix Composite Materials* (ASTM International, West Conshohocken, PA)
[16] Santulli C and V. Arumugam S B K 2011 Effect of Fiber Orientation in Uni-Directional Glass Epoxy Laminate Using Acoustic Emission Monitoring 金属学报英文版 **24** 351–64

IOP Conf. Series: Materials Science and Engineering **726** (2020) 012008 doi:10.1088/1757-899X/726/1/012008

The physical – mechanical properties of low-density polyethylene films

L Markovičová, V Zatkalíková, T Kojnoková, D Gaňa, T Liptáková

University of Žilina, Faculty of Mechanical Engineering, Department of Material Engineering, Univerzitná 8215/1, 01026 Žilina, Slovakia

lenka.markovicova@fstroj.uniza.sk

Abstract. Polyethylene (PE) is the most commonly used polymer. Most of the polymers are processed by the packaging sector. PE products are the most recycled plastic due to the short life of packaging materials. The experiment is based on real issues in the recycling of plastic waste in a world that is increasingly up to date. One way of recycling it is to produce returnable plastic bottles that can be used up to 50 times. The low-density polyethylene (LDPE) film is applied to the plastic bottles before filling with a specific medium. After the contents of the bottle have been consumed, the LDPE film is stripped. Subsequently, a new recyclable film is applied to the bottles used and the process is repeated. Changes in the properties of LDPE films applied to these returnable bottles have been evaluated by several tests (tensile test and water absorption test). The tests were performed on unused LDPE films and on used LDPE films, which were filled with the medium for 3 months. The fill media was chosen from commonly available goods.

1. Introduction

Polyethylene is considered a plastic material with long-term mechanical properties, easy treatability and weldability. These properties decided for using of low density polyethylene (LDPE) films for protection of plastic containers. Application of the LDPE films as wrapping on container transported often aggressive media makes possible their cyclic using.

Nowadays, the container is determined only for one usage and it is liquidate after usage. By application of the wrap foils the economical as well as environmental effect can be achieved because decreasing of plastic waste is very urgent in this era. The eventuality of multiple times using of plastic container the LDPE films must have optimal properties to be able protect containers against contamination [1].

Degradation of polymers according to effecting factors (temperature, mechanical loading, present microorganism, chemicals etc.) can be shown like thermo-degradation, biodegradation, oxidation, mechanical degradation etc. [2]. The processes of the forming have negligible effect but they can evoke structural changes with influence on long-term behaviour [3, 4]. The effect of elevated temperature during forming can be observed because the assumption of heat increased kinetic energy of macromolecules can overcome attachment forces among polymers and they can behave like liquids. The polymers structure is changed by various ways (breaking or crosslink of polymer chain [5-7]. The goal of our work is to compare the mentioned properties of the original LDPE films and after their application to container.

2. Experimental material

The investigated foils are made from low-density polyethylene of the thickness 50 ± 2 μm welded in longitudinal direction. Low density is caused by high number of the side chain CH_3. Crystallinity is about 40 – 50 %. Experiments are focused on investigation of mechanical properties of the LDPE foils after their exposure in the coca cola marked C (acid environment, pH = 2.23), disinfectant marked S (the main component is the sodium hypochlorite, pH = 13.37) and tenside cleaner marked SR (pH = 3.35). Foil samples were exposed for 3 months in all the three media. After the exposure, all the evaluated parameters were compared with the unaffected foil (O LDPE).

3. Experimental methods and results

The tensile strength characteristics of the PE are sensitive to structural changes. This is because the crystalline regions significantly affect the modulus. The other effect observed is that of strengthening due to the molecular orientation imparted during the film blowing. This is because of the molecular grade tensile properties are higher in the direction of the covalently bonded carbon – carbon chain than in the transverse direction, which is dominated by the much weaker van der Waals bonds [8-10]. The effect of various exposure environment and exposure time on the LDPE mechanical properties foils in the longitudinal and transverse directions were examined. The tensile strength at break was determined according to STN EN ISO 527 [11], with a jaw speed of 50 $mm.min^{-1}$. In each direction (longitudinal – L, transverse – T) 5 samples were tested, the results being the average of 5 measurements. The tensile strength after 3 months of exposure increased in all the environments compared to the tensile strength of the original sample (figure 1). On the contrary, the disinfectant caused the tensile strength increase approximately 2.5 times compared to the original foil. Figure 2 shows the change in tensile strength at break depending on the environment and exposure time.

The water absorption test for plastics is carried out in cold (method A) and in boiling water (method C). The method consists in determining the weight gain of test pieces immersed in cold and boiling water over a prescribed time. The test specimens should have the same shape and dimensions. At least 3 immersion test bodies shall be tested the prescribed time into the water container. After the time has elapsed, they are removed from the water, dried and weighed to the nearest 1 mg (ISO 640112) [12].

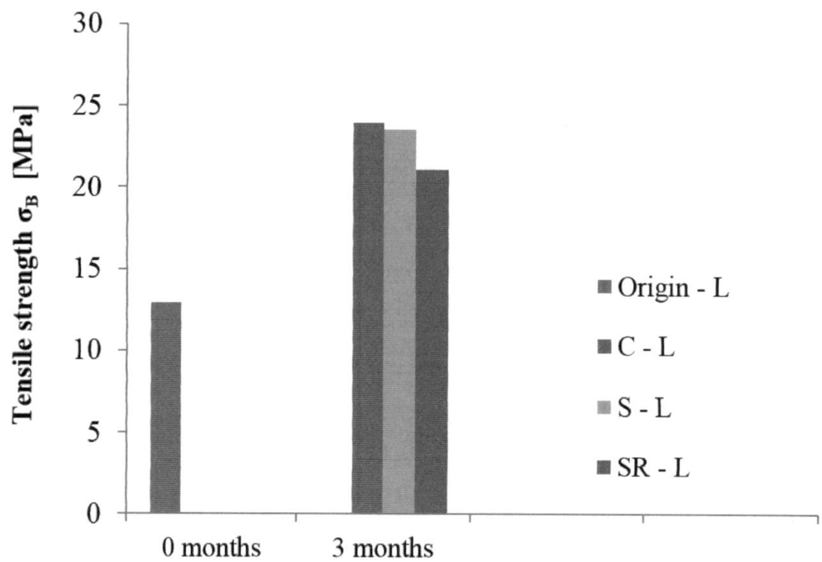

Figure 1. Dependence of the breaking strength of the LDPE film on exposure environment and exposure time in longitudinal direction.

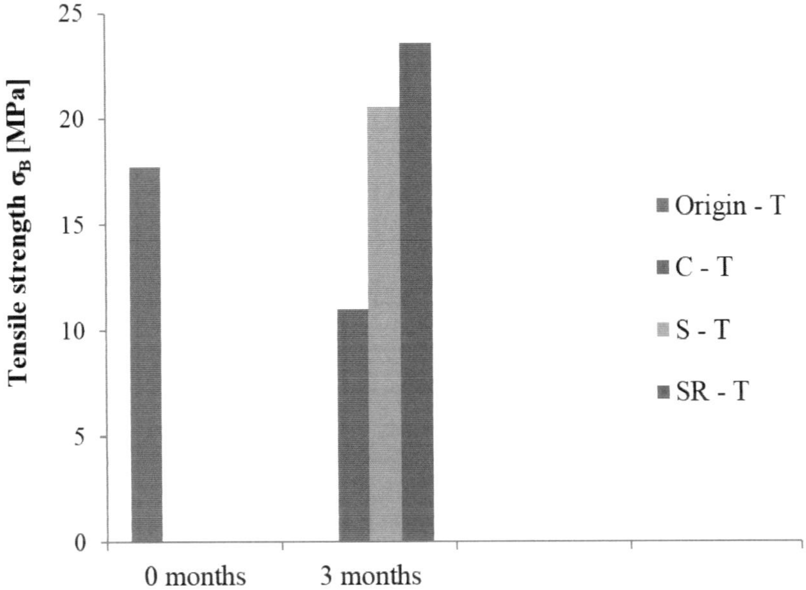

Figure 2. Dependence of the breaking strength of LDPE film from exposure environment and exposure time in transverse direction.

The amount of water absorbed by the test specimen in mg is calculated using method A (X_1) and C (X_2) according to formula 1:

$$X_{1,2} = m_2 - m_1 \quad (mg) \tag{1}$$

Amount of water absorbed by the test piece per unit of surface in mg.cm^{-2} is calculated using method A (X_5) and method C (X_6) using formula 2:

$$X_{5,6} = \frac{m_2 - m_1}{A} \quad (mg.cm^{-2}) \tag{2}$$

The amount of water absorbed by the test specimen in % is calculated using method A (X_9) and method C (X_{10}), using formula 3:

$$X_{9,10} = \frac{m_2 - m_1}{m_1} \cdot 100 \quad (\%) \tag{3}$$

Mean values of water absorption for individual methods of water absorption determination are given in table 1. In cold water at 23 °C, the LDPE film absorbed some of the fluid based on the results, increasing its original weight. In cold water for 24 hours, there was in all cases a slight increase in the weight of the LDPE film by the selected media after 3 months. This represents in % the water absorption of the LDPE film in cold water from 0.2 to 1 %.

In boiling water at 100 °C, the weight of the LDPE film decreased to 0.6 % in the case of Savo compared to the original LDPE.

Table 1. Absorbency of origin O LDPE film and LDPE film affected by media.

Sample	Sample area (cm²)	Sample weight (mg) Before absorption m_1	After absorption m_2	$X_{1,2}$ (mg)	$X_{5,6}$ (mg.cm^{-2})	$X_{9,10}$ (%)
			Cold water			
O LD-PE film	2.83	12.60	12.67	0.067	0.024	0.531
Savo	1.43	7.62	7.66	0.040	0.030	0.570
Saponate	1.52	8.22	8.29	0.077	0.048	0.945
Coca-Cola	1.87	8.72	8.74	0.023	0.012	0.271
			Boiling water			
O LD-PE film	2.82	12.40	12.48	0.260	0.056	0.705
Savo	1.34	7.14	7.11	-0.027	-0.019	-0.570
Saponate	1.78	7.81	7.88	0.070	0.042	0.947
Coca-Cola	1.64	8.29	8.36	0.067	0.042	0.813

The absorbency values of LDPE film in cold and boiling water expressed in mg are shown in figure 3. Higher water absorption in boiling water compared to cold water had O LDPE film and LDPE film affected by Coca-Cola. LDPE was influenced by Savo and Saponate to have a higher water absorption in cold water than boiling water. O LDPE film in boiling water has the highest absorbency.

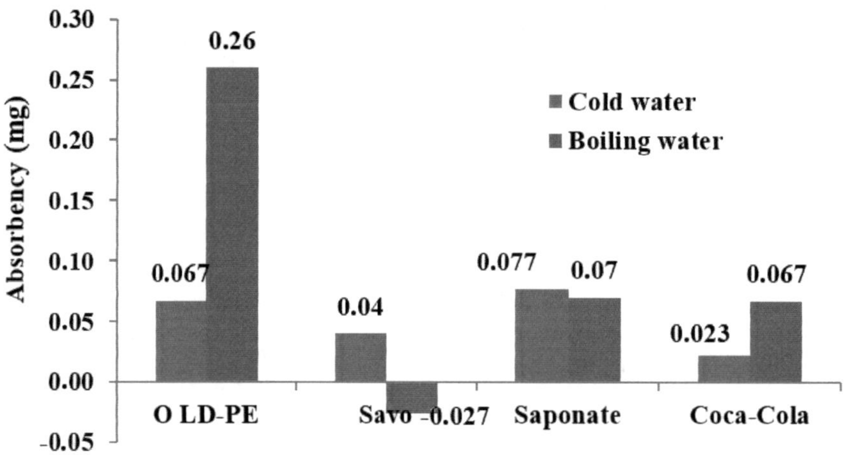

Figure 3. Absorbency of O LDPE film and LDPE film influenced by medium in mg.

4. Conclusions

Based on experiments performed on LDPE films after exposure in 3 different environments for 3 months, we can conclude:

- Evaluation of the tensile strength test results are more complicated because they are influenced by the loading direction (longitudinal, transversal), as well as by regularity of the surface attack of media.
- The tensile strength of the LDPE foils increased in all the media already after 3 month exposure in both tested directions. It can be caused by physical effect of the liquid media which penetrate to intermolecular area.

- The absorption of samples exposed to all three environments resulted in increased sample weight. Only in the case of Savo there was a weight loss, which could be due to the chlorine content of Savo, which during exposure was poorly bound to the polyethylene chains and was released by the elevated temperature in the absorbency test.
- Based on the results, the water absorption of the LD-PE film in the water absorption test is up to 1 % in cold water and up to 0.6 % in boiling water.

Acknowledgement
The research was supported partially by Scientific Grant Agency of Ministry of Education, Science and Sport of Slovak Republic and Slovak Academy of Science, grant KEGA No. 049ŽU-4/2017 and grant KEGA No. 013ŽU-4/2019.

This work was also supported under the project of Operational Programme Research and Innovation: Research and development activities of the University of Zilina in the Industry of 21st century in the field of materials and nanotechnologies, No. 313011T426. The project is co-funding by European Regional Development Fund.

References

[1] Achilias D S, Roupakias C, Megalokonomos M, Lappas A A and Antonakou E V 2007 Chemical recycling of plastic wastes made from polyethylene (LDPE and HDPE) and polypropylene (PP) *J. Hazard. Mater.* **149** 536-542
[2] Ducháček V 1999 *Polymery – výroba, vlastnosti, zpracování, použití* (Praha: VŠCHT) p. 190 ISBN 80-7080-617-6
[3] Capone C, Di Landro L, Inzoli F, Penco M and Sartore L 2007 Thermal and mechanical degradation during polymer extrusion processing *Polym. Eng. Sci.* **47** 1813-1819
[4] McMurry J 2004 *Organic chemistry* (Brooks/Cole: Thomson Learning Company) ISBN 0534389996
[5] Hassan A, Yahya R, Yahaya A, Tahir A R M and Hornsby P R 2004 Tensile, Impact and Fiber Length Properties of Injection-Molded Short and Long Glass Fiber-Reinforced Polyamide 6,6 Composites *J. Reinf. Plast. Comp.* **23** 969-986
[6] Allen N S 2010 *Photochemistry and Photophysics of Polymer Materials* (Hoboken, New Jersey Canada: John Wiley&Sons, Inc.) ISBN 978-0-470-13796-3
[7] Allen N S, Edge M 1992 *Fundamentals of Polymer Degradation and Stabilization* (Heidelberg: Springer Science & Business Media) ISBN 1-85166-773-3
[8] Powell P C 1977 *The selection and use of thermoplastics* (London: Oxford University Press)
[9] Gruenwald G 1994 *How Structure Determines Properties* (Munich: Hanser Publisher)
[10] Vishu S 1984 *Handbook of Plastic Testing Technology* (New York: Wiley)
[11] Standards STN EN ISO 527 Plastics. Determination of tensile properties.
[12] Standards EN ISO 62 (640112) Plastics. Determination of water absorption in water.

IOP Conf. Series: Materials Science and Engineering **726** (2020) 012009 doi:10.1088/1757-899X/726/1/012009

Construction of magnetorheological device for finishing of non-metallic materials

R Milde, O Bílek, M Sedlačík, M Kovařík

Tomas Bata University in Zlín, Faculty of Technology, Vavreckova 275, 760 01 Zlín, Czech Republic

rmilde@utb.cz

Abstract. The paper deals with nano-finishing processes and nano-finishing processes in the presence of an external magnetic field. There is a construction of magnetorheological device for finishing of non-metallic materials and subsequent verification of the functionality on the change in micro-roughness of the PA-6 material.

1. Introduction

Magnetorheological polishing is a finishing method where a polishing effect is achieved in the presence of a magnetic field and a magnetorheological (MR) suspension generally consisting of magnetic microparticles, e. g. carbonyl iron powder (CIP), abrasive grains, carrier fluid, and other additives [1-3]. This composition gives the system specific field-responsive properties. The field-responsive properties of the MR suspension increase with increasing external magnetic field. When the relative movement between the MR suspension and the workpiece occurs, the normal force acts on the abrasive grains and the reaction shear forces the material in the form of micro- or nano-chips [4-8]. These chips mix with MR suspension and reduce machining efficiency. This results in wear, so it is necessary to replace the used suspension at regular intervals in order to minimize the decreasing MR effect as a negative phenomenon [9-10].

2. Construction of the device

The proposed device is designed only for planar surfaces. The repeatability of the speed control, the direction of rotation, the parallelism between the polishing surface and the polished surface must be ensured. Only non-metallic materials can be polished using the developed machine. The individual constructional elements of MR device are following.

2.1. Magnet drive design

The first step in designing the construction of MR device for finishing of non-metallic materials was to determine the magnet drive. To achieve the torque transmission, toothed belt transmissions have been chosen to guarantee high torque transmission accuracy. Their advantages include (i) a form-fit in which they work without slip, (ii) can carry heavy loads with low space requirements and (iii) are quiet. The bearings do not require pre-stressing, which reduces their stressing. They do not require maintenance and lubrication. Among the disadvantages of toothed belts is the susceptibility to lateral slip of the belt, therefore at least one pulley is provided with a flange. In the vertical position of the pulley axes, both pulleys must be provided with guide flanges. In order to ensure a quiet and calm running of the belts, it is necessary to maintain a sufficient parallelism of the pulley axes. The average permissible value is ± 0.25 °.

2.2. Determination of the carrier drive

The Z axis, which is equipped with a position indicator with a resolution of 0.01 mm, was chosen to ensure repeatability of the distance between the carrier and the workpiece. The material of the shaft was chosen low carbon steel W.Nr. 1.0038. The trapezoidal screw Tr 20 × 4 is used to convert the rotary motion into a sliding motion. The cylinder on which the magnet is attached has a thickness of 10 mm as seen in figure 1. The maximum stroke of the axis is 212 mm.

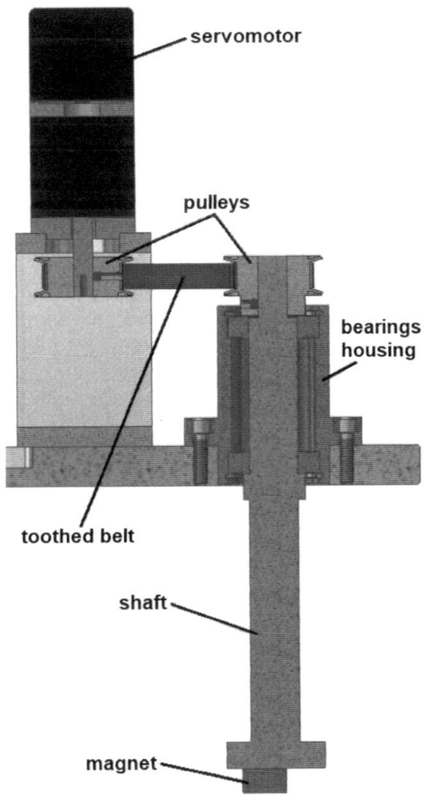

Figure 1. Scheme of the magnet drive. Check calculation

Ball bearings were chosen to accommodate the rotary parts. These bearings are designed for small and medium loads and can carry a part of the axial load in addition to the radial load. High stiffness bearings, such as ball or roller bearings, cannot compensate for misalignment. That compensate for only a very small misalignment when there is almost no load. The most stressed part is the plate that carries the drives. For this board the calculation was performed in the CATIA program as shown in figure 2.

2.3. Description of output technical properties

The device for polishing non-metallic materials is designed for polishing any flat surface with a maximum dimension of 40 x 40 mm. The size of the gap between the material and the magnet can be controlled by a mobile mechanism in the range of 0 – 160 mm with an accuracy of 0.1 mm. The rigidity and vibration damping of the entire system ensure the specific arrangement of the plates and the position of the motors relative to the polishing axis (figure 2).

IOP Conf. Series: Materials Science and Engineering **726** (2020) 012009 doi:10.1088/1757-899X/726/1/012009

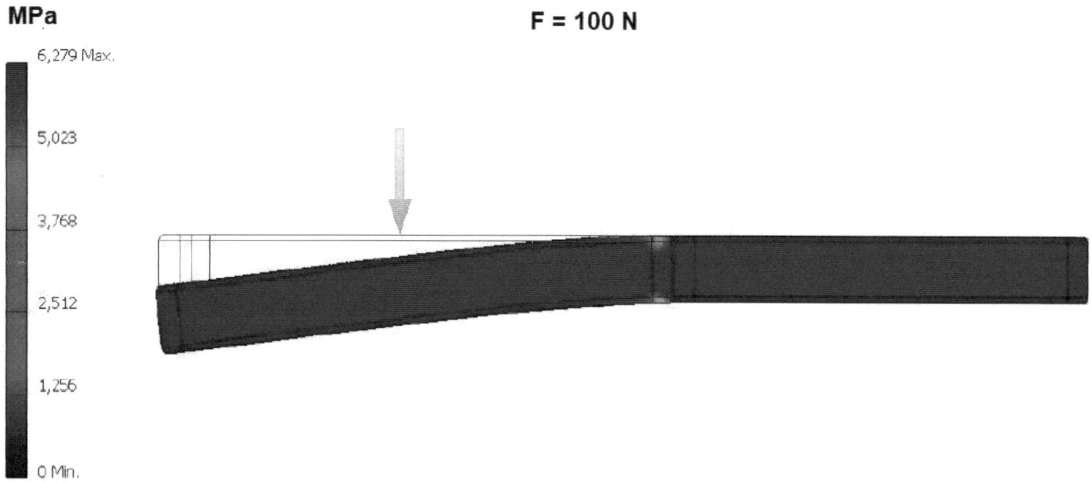

Figure 2. Verification of the mounting plate.

The motors also ensure speed control and smoothness of the entire polishing process. The whole device is low maintenance, which complements the overall environmental and energy-saving process in general. With the above advantages, a better surface finish can be achieved than with the conventional lapping method. A mobile mechanism is mounted on the base, which is primarily designed to adjust the distance between the magnet and the sample (including the MR suspension). It also serves for the vertical movement of the main part of the mechanism. It contains two 100W motors designed to rotate the magnet placed under the cover. Belts connecting the motors to the shaft mechanism allow this movement as seen in figure 2. The sample is located in a vice, which ensures the flatness of polishing between the magnet and the sample.

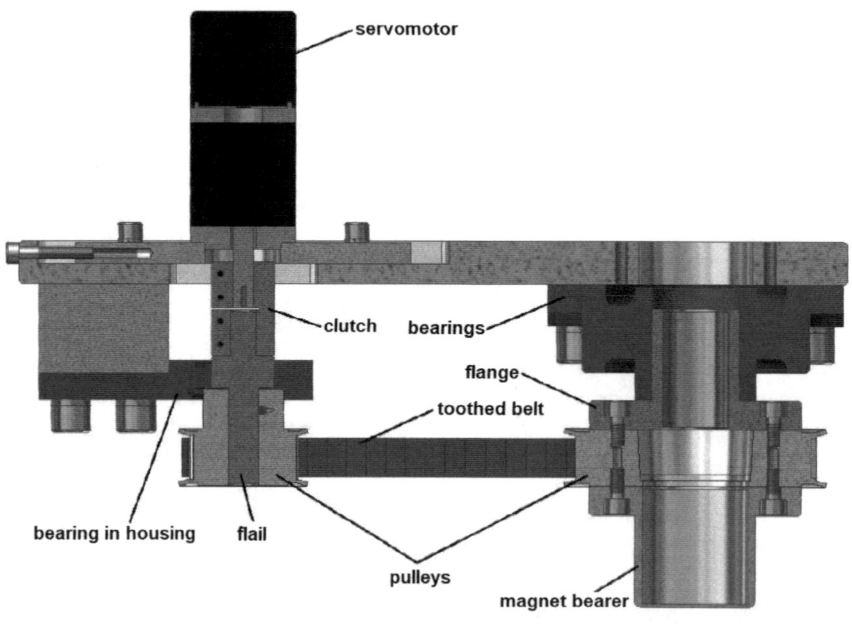

Figure 3. Scheme of the carrier drive.

3. Verification of the functionality of the device

The functionality of the device was verified on PA-6 RAVAMID B, which was dried at 80 °C for 12 hrs before testing. The test samples (20 × 20 mm) were injected on an Arburg Allrounder 470 E Golden Electric. The mold temperature was 50 °C, the injection pressure was 1 200 bar, the pressure time was 15 s, and the entire cycle lasted 48.5 s. Each test sample was then cast into GAFORM D30 resin and cured for 1 h. Subsequently, 10 samples were subjected to a polishing test, where the surface quality was evaluated using a Mitutoyo SJ-410 roughness tester according to ISO1997 at regular intervals of 5 min until there was not significant improvement in the resulting surface. The polishing medium used was MR suspension containing CIP (60 wt.%), water-based magnetic fluid (30 wt.%), and alumina (10 wt.%). Figure 4 represents the polishing curve expressed as dependence of surface roughness, *Ra*, on the time of polishing. Evidently, the equilibrium of $Ra \approx 0.06$ μm was obtained after 25 min of polishing using the MR suspension.

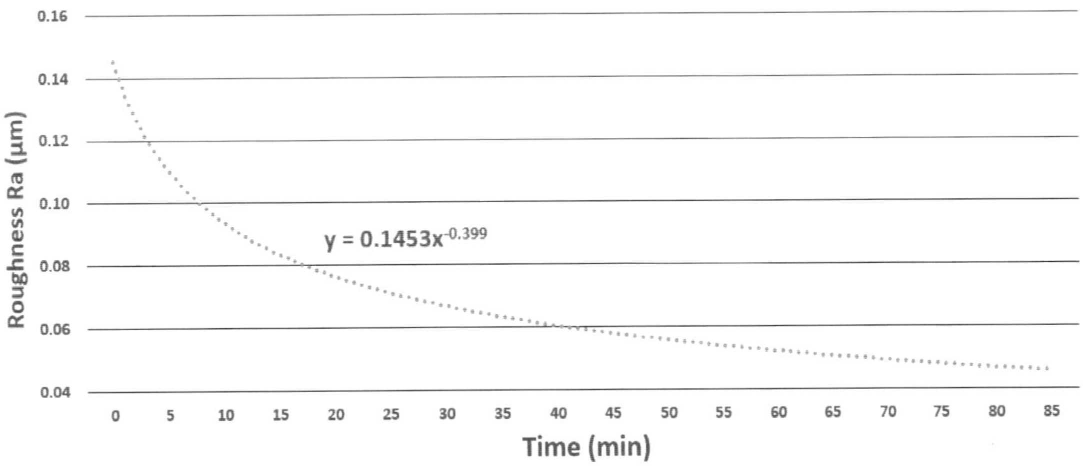

Figure 4. Polishing curve using MR suspension.

4. Conclusion

Polishing of ten PA-6 samples for 1.5 hrs in the presence of the MR suspension has proven the functionality of the device. All the samples showed a surface improvement of at least 34 %. Furthermore, it has been shown that the proposed suspension has polishing properties for polishing non-metallic surfaces and will be subject to further investigation.

Acknowledgement

This work and the project is realized with the financial support of the internal grant of Tomas Bata University in Zlín No. IGA/FT/2019/001 funded from the resources of specific university research.

References

[1] Jain V K 2013 *Micromanufacturing processes* (Boca Raton, FL: Taylor & Francis)
[2] Bednarek S 2001 The giant linear magnetostriction in elastic ferromagnetic composites within a porous matrix *J. Magn. Magn. Mater.* **301** 200–7
[3] Laun H M, Gabriel C and Kieburg C 2010 Magnetorheological fluid in oscillatory shear and parameterization with regard to MR device properties *J. Intell. Mater. Syst. Struct.* **21** 1479-89

[4] Jiang W, Zhang Y, Xuan S, Guo C and Gong X 2011 Dimorphic magnetorheological fluid with improved rheological properties *J. Magn. Magn. Mater.* **323** 3246–50
[5] Abramchuk S S, Grishin D A, Kramarenko E Y, Stepanov G V and Khokhlov A R. 2006 Effect of a homogeneous magnetic field on the mechanical behavior of soft magnetic elastomers under compression *Polym. Sci. Ser. A.* **48** 138–45
[6] Wang Y, Wu Y and Mitsuyoshi N 2016 A novel magnetic field-assisted polishing method using magnetic compound slurry and its performance in mirror surface finishing of miniature V-grooves *Aip. Adv.* **6** 56602
[7] Kim B C, Chung J H, Cho M W, Ha S J and Yoon G S 2018 Magnetorheological fluid polishing using an electromagnet with straight pole-piece for improving material removal rate *J. Mech. Sci. Technol.* **32** 3345–50
[8] Guo H and Wu Y 2016 Ultrafine polishing of optical polymer with zirconia-coated carbonyl-iron particle-based magnetic compound fluid slurry *Int. J. Adv. Manuf. Technol.* **85** 253–61
[9] Lee J W, Hong K P, Cho M W, Kwon S H and Choi H J 2015 Polishing characteristics of optical glass using PMMA-coated carbonyl-iron-based magnetorheological fluid *Smart. Mater. Struct.* **24** 65002
[10] Lee J W, Hong K P, Kwon S H, Choi H J and Cho M W 2017 Suspension rheology and magneto-rheological finishing characteristics of biopolymer-coated carbonyliron particles *Ind. Eng. Chem. Res.* **56** 2416–24

IOP Conf. Series: Materials Science and Engineering **726** (2020) 012010 doi:10.1088/1757-899X/726/1/012010

On the problem of processing data from a cross-sectional image of grained structure

S Minárik and M Martinkovič

Slovak University of Technology in Bratislava, Faculty of Materials Science and Technology in Trnava, Ulica Jána Bottu č. 2781/25, 917 24 Trnava, Slovak Republic

stanislav.minarik@stuba.sk

Abstract. The anisotropy of a grained structure is a consequence of non-uniform distribution of spatial orientation of grains. Conventional stereological methods allow determine spatial distribution of grain orientation from lines distribution observed on structure cross sections. The distribution of grain`s surface orientation in space is determined by measuring the total length of cross sections on a cutting plane. In this paper, we indicate partial but significant problem of stereology. A rectangular model of grained structure is considered. Then the kernel function, which connects the probability density function (distribution) of lengths of rectangles intersections visible on observation planes with size of rectangles sides, is calculated.

1. Introduction

During last decades a lot of sophisticated methods have been developed to study structures by 2D imaging. These methods can significantly help in precise investigation of grained structures. On the other hand these methods have limitations. They are neither always economically efficient nor can be applied for different situations. Mentioned methods certainly cannot be considered fast and cheap and therefore it makes sense to look for simpler alternatives. Stereological methods are promising, very simply and cheap for this application.

A growing literature is demonstrating that problems of geometrical orientation of structure and structural units distortion induced by deformation play important role in different areas. As it is clear from previous work, there are a few ways how to evaluate the grain orientation experimentally [1-6]. However, the most suitable way in practice is the scalar measurement of anisotropy based on stereological principle enabling to determine the degree of grains orientation. Using stereological method an oriented test lines enables to determine the degree of grains orientation in any area of polycrystalline pieces [7-10]. In addition, we suggest that results of such measurement, i.e. degree of grains orientation in volume of plastically deformed structure can be used to evaluation of local deformation parameters. In that case, it is sufficient measure the degree of grains orientation to obtain a value of local strain. Our work offers some details of statistical view of grained structures analysis using stereological concept.

2. The problem formulation

Let's look the grained structure as a set of grain boundaries. Consider rectangle model of grain boundary in bulk grained structure with shape of cube.

IOP Conf. Series: Materials Science and Engineering **726** (2020) 012010 doi:10.1088/1757-899X/726/1/012010

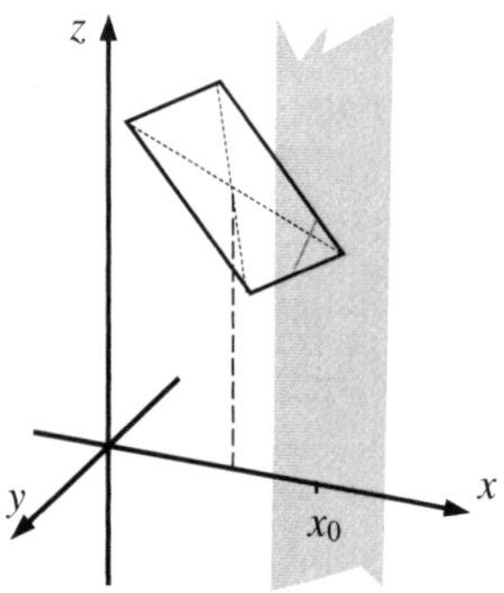

Figure 1. A view on intersection of randomly placed rectangle with a plane $x = x_0$.

Arbitrary placed rectangle intersects the cross-sectional plane $x = x_0$ (Figure 1) if

$$x_0 - x \le \frac{L}{2}|cos\,\alpha| \quad \ldots \quad \alpha \in \langle 0, 2\pi \rangle,$$

where

$$L = a|cos(\beta)| + b|sin(\beta)| \quad \ldots \quad , \beta \in \langle 0, 2\pi \rangle,$$

a and b are length of rectangle sides. In the limit case it holds $L_0 = u\,sin(\varphi)$, where $u = \sqrt{a^2 + b^2}$.

Then from the integral geometry technique results that the probability that the rectangle intersects the cross-sectional plane can be determined as

$$P = \frac{M}{M_0}$$

where

$$M_0 = (2\pi.2\pi.d) = (4\pi^2 d) \text{ and } M = \frac{1}{2}\int_0^{2\pi}\int_0^{2\pi}\{a|cos(\beta)| + b|sin(\beta)|\}|cos\,\alpha|\,d\beta\, d\alpha = 8(a+b).$$

If the sample has the shape of cube with the side d, mentioned probability can be calculated by simple formula

$$P = \frac{2(a+b)}{\pi^2 d}.$$

However if the rectangle intersects the cross-sectional plane, there can be various length of the intersection line. If the intersection shown in Figurer 1 is allready occured we can expect next possile configurations of intersection line determined by points C_1 and C_2 positioned in the rectangle sides

$C_1 = [x_1 ; y_1]$ $C_2 = [x_2 ; y_2]$

1) $P(x_1 \in \langle 0,a \rangle \wedge y_1 = 0 \wedge x_2 \in \langle 0,a \rangle \wedge y_2 = b)$
2) $P(x_1 \in \langle 0,a \rangle \wedge y_1 = 0 \wedge x_2 = 0 \wedge y_2 \in \langle 0,b \rangle)$
3) $P(x_1 \in \langle 0,a \rangle \wedge y_1 = 0 \wedge x_2 = a \wedge y_2 \in \langle 0,b \rangle)$
4) $P(x_1 = 0 \wedge y_1 \in \langle 0,b \rangle \wedge x_2 \in \langle 0,a \rangle \wedge y_2 = b)$
5) $P(x_1 = a \wedge y_1 \in \langle 0,b \rangle \wedge x_2 \in \langle 0,a \rangle \wedge y_2 = b)$
6) $P(x_1 = 0 \wedge y_1 \in \langle 0,b \rangle \wedge x_2 = a \wedge y_2 \in \langle 0,b \rangle)$

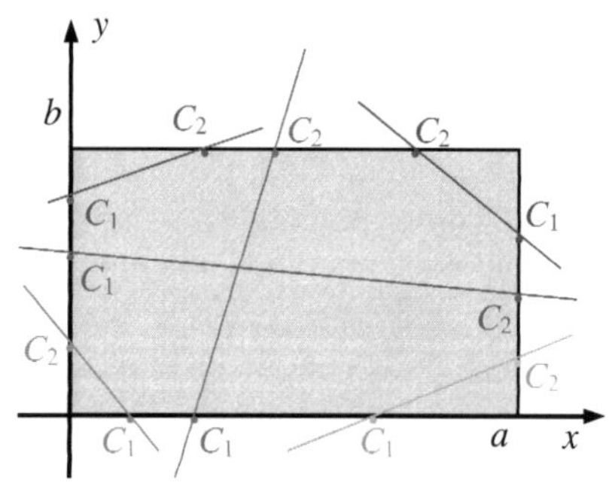

Figure 2. A view on possible intersection points C_1 and C_2 of rectangle with a plane corresponding

The question is, what is probability, that length of the intersection line (i. e. distance of points C_1 and C_2) is smaller then a value ξ. All situations corresponding with configurations 1) to 6) shown in Figure 2 must be considered, i. e.

1) $P(A_1) = P\left(\sqrt{(x_1 - x_2)^2 + b^2} < \xi\right)$, 2) $P(A_2) = P\left(\sqrt{x_1^2 + y_1^2} < \xi\right)$,

3) $P(A_3) = P\left(\sqrt{(x_1 - a)^2 + y_2^2} < \xi\right)$, 4) $P(A_4) = P\left(\sqrt{x_2^2 + (y_1 - b)^2} < \xi\right)$,

5) $P(A_5) = P\left(\sqrt{(x_2 - a)^2 + (y_2 - b)^2} < \xi\right)$, 6) $P(A_6) = P\left(\sqrt{a^2 + (y_1 - y_2)^2} < \xi\right)$ ... $\xi \in \left\langle 0, \sqrt{a^2 + b^2} \right\rangle$.

It is clear, that

$$P(\Lambda_2)=P(\Lambda_3)=P(\Lambda_4)=P(\Lambda_5)=\Pi(\gamma) \quad \ldots \quad \Pi(\gamma)=\Pi\left(\sqrt{x^2+y^2}<\xi\right)$$

and the probability is

$$P(\xi)=C_1P(\Lambda_1)+C_2\Pi(\gamma)+C_3P(\Lambda_6),$$

where C_1, C_2 and C_3 are constants resulting from normalization of distribution function $f(\xi)$ determined by derivative of P.

3. The length of line visible in the cross-section

We briefly demonstrate analysis of configuration considering 1) in which $\xi > b$. Let a is the length of the interval at which the cross-section points in structure is carrying out. Consider two different randomly chosen cross-section points C_1, C_2 formed in the sides with length a. Let x_1 and x_2 denote positions where cross-section C_1 and C_2 are formed. Then the length of the cross-section line is valuable as

$$L=\sqrt{(x_1-x_2)^2+b^2}$$

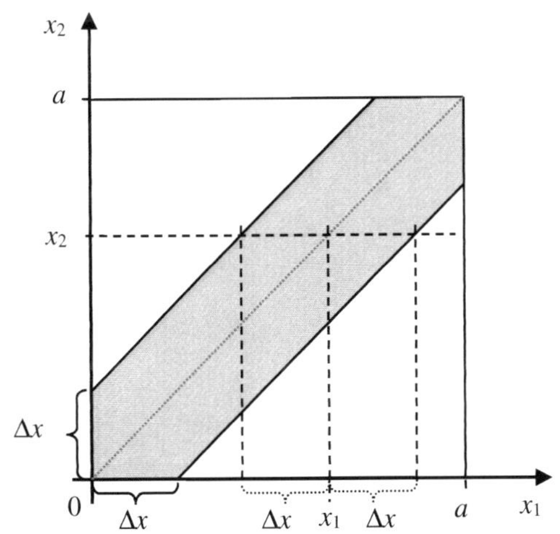

Figure 3. Scheme illustrating a pair of cross-section points A, B phenomena in time domain. The grey area in diagram corresponds to the case when $\Delta x < \sqrt{\xi^2-b^2}$.

Actually, we are interested in the probability $P(\Lambda_1)$ that the interval L is shorten then arbitrarily chosen value ξ, i.e. $P(\Lambda_1)$ denotes the expectation of the cross-section length between C_1 and C_2 is not longer then ξ

$$\sqrt{(x_1-x_2)^2+b^2}<\xi \ldots \Delta x=|x_1-x_2|<\sqrt{\xi^2-b^2}.$$

Problem of the probability $P(\Lambda_1)$ can be solved by computations with conditional probability. It is, however, more instructive to solve it by another method, one that minimizes the amount of computation and maximizes the role of probabilistic reasoning. Application of integral geometry technique is appropriate for the probability $P(\Lambda_1)$ evaluation in this case.

Like combinatorial enumeration, where sequences of objects bearing a common feature are unified by the idea of a generating function, integral geometry studies sets of geometric objects bearing a common feature, which are unified by the idea of an invariant measure. The basic idea is extremely simple. With integral geometry, we search for the likelihood that we will "hit" a particular area of a figure. To calculate the probability $P(\Lambda_1)$, we will need to find 2D areas of the shapes involved in the problem. We will need to know the total area, which means the biggest area in the diagram, like the entire dartboard. We will also need to know the desired area, which is the part we are trying to "hit".

2D metric space can be identified to explain interval scaling of the cross-section points C_1, C_2 formation. We can define axes that allow to illustrate positions where both of C_1 and C_2 are created (Figure 3). Intervals with lengths a can be easily found on both of these axes. Every point on the surface of square shown in 3 represent possible events of the C_1, C_2 positions. Then the number of all possible events is represented by area of this square S_0. Number of favourable formation events (i. e. events when the time interval Δx is shorten then $\sqrt{\xi^2 - b^2}$) is represented by the grey area S shown in Figure 1. This is the desired stands for the area that we want to "hit". Once we can easy calculate both of these areas and the formula for the probability $P(\Lambda_1)$ calculation is simply

$$P(\Lambda_1) = 1 - \left(\frac{a - \sqrt{\xi^2 - b^2}}{a} \right)^2 .$$

The configuration 6) ($\xi > a$) can be analyzed by the same way and the result is analogous

$$P(\Lambda_6) = 1 - \left(1 - \frac{\sqrt{\xi^2 - a^2}}{b} \right)^2 .$$

If the configuration corresponding to the probability $\Pi(\gamma)$ is analyzed using integral geometry technique described above, obtained result shows that three situations must be considered $\xi < b$, $b < \xi < a$ and $a < \xi$

I. If $\xi < b$ then $P(\xi) = C_1 \pi \xi^2$

II. If $b < \xi < a$ then $P(\xi) = C_1 \{1 - \varepsilon_1\} \pi \xi^2 + C_2 \left\{ 1 - \left(1 - \frac{\sqrt{\xi^2 - b^2}}{a} \right)^2 \right\}$

III. If $a < \xi$ then $P(\xi) = C_1 \{1 - \varepsilon_1 - \varepsilon_2\} \pi \xi^2 + C_2 \left\{ 1 - \left(1 - \frac{\sqrt{\xi^2 - b^2}}{a} \right)^2 \right\} + C_3 \left\{ 1 - \left(1 - \frac{\sqrt{\xi^2 - a^2}}{b} \right)^2 \right\}$

where

$$\varepsilon_1 = \varepsilon_1(\xi) = \frac{2}{\pi} \left\{ arcsin\left(\sqrt{1 - \left(\frac{b}{\xi} \right)^2} \right) - \frac{b}{\xi} \sqrt{1 - \left(\frac{b}{\xi} \right)^2} \right\} , \quad \varepsilon_2 = \varepsilon_2(\xi) = \frac{2}{\pi} \left\{ \frac{\pi}{2} - arcsin\left(\frac{a}{\xi} \right) - \frac{a}{\xi} \sqrt{1 - \left(\frac{a}{\xi} \right)^2} \right\}$$

It is evident, that derivative of $P(\xi)$ shows singularities for $\xi = a$ and $\xi = b$. To avoid this fact the generalized function describing the distribution of length of line elements visible in the cross-section was evaluated numerically as $f(\xi) = P(<\xi, \xi + d\xi>)$. Calculated data for the $a = 10$ and $b = 5$ is shown in Figure 4.

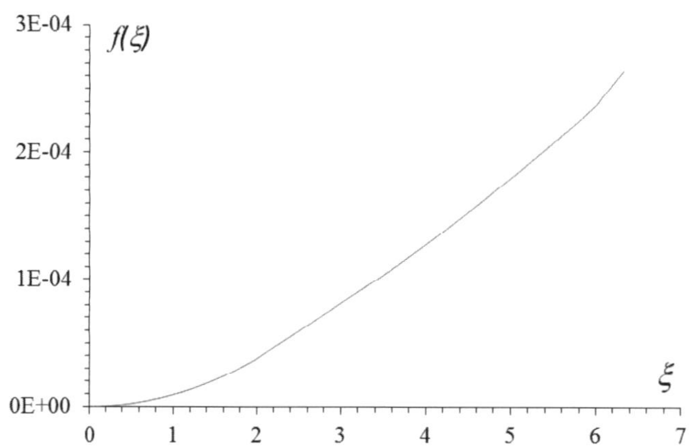

Figure 4. Distribution function for length of the line element visible in cross-section.

4. Conclusion

It is clear that real grain boundaries haven`t rectangular form. Boundaries of various shape are in the grained structure. Therefore obtained analytical formulas must be generalized to suitable forms. However our result document that there is a distribution of length of line elements visible in the cross-section that would be considered when the conventional stereological techniques are applied.

Finally, results from all conventional 2D imaging techniques using a cross-sectional image of the structure would be analyzed with the consideration of this fact.

Acknowledgment

This work was supported by the Slovak Research and Development Agency under the contract No. APVV-15-0319 and by the VEGA Grant No. 1/0122/16 of the Grant Agency of the Slovak Republic Ministry of Education.

References

[1] Sutton M A, Orteu J-J and Schreier H W 2009 *Image Correlation for Shape, Motion and Deformation Measurements* (Hardcover ISBN 978-0-387-78746-6)

[2] Sack R A 1961 Indirect Evaluation of Orientation in Polycrystalline Materials *J. Polym. Sci.* **51** 543-560

[3] Keating T J, Wolf P R and Scarpace F L 1975 An Improved Method of Digital Image Correlation *Photogramm. Eng. Rem. S.* **41** 993–1002

[4] Masteghin M G and Orlandi M O 2016 Grain-Boundary Resistance and Nonlinear Coefficient Correlation for SnO_2-Based Varistors *Mat. Res.* **19** (São Carlos Epub 26 Sep 2016)

[5] Bunkholt S and Knut M and Nes E 2014 Subgrain Structures Characterized by Electron Backscatter Diffraction *Mater. Sci. Forum* **794-796** 3-8

[6] Zou H F and Zhang Z F 2010 Application of electron backscatter diffraction to the study on orientation distribution of intermetallic compounds at heterogeneous interfaces (Sn/Ag and Sn/Cu) *J. App. Phys.* **108** 103518

[7] Saltykov S A 1970 *Stereometric metallography* (Metallurgia, Moscow, in Russian)

[8] Lewis H D, Walters K L and Johnson K A 1973 Particle size distribution by area analysis: Modifications and extensions of the Saltykov method *Metallography* **6** 93–101

[9] Xu Y H and Pitot H C 2003 An improved stereologic method for three-dimensional estimation of particle size distribution from observations in two dimensions and its application *Comput. Methods Programs Biomed.* **72** 1-20

[10] Rayaprolu D B and Jaffrey D 1982 Comparison of discrete particle sectioning correction methods based on section diameter and area *Metallography* **15** 193–202

IOP Conf. Series: Materials Science and Engineering **726** (2020) 012011 doi:10.1088/1757-899X/726/1/012011

Pulse measurements of small area thin film μc-Si:H/ZnO:B photodiodes

Z Remes and J Stuchlik

FZU - Institute of Physics of the Czech Academy of Sciences, Na Slovance 2, Prague, CZ-18221, Czech Republic

remes@fzu.cz

Abstract. We introduce a triggered optoelectronic system operating in a pulse mode in the near infrared and visible spectral range 0.75–3 eV. The system measures current-voltage (*I-V*) characteristics in dark and under visible light illumination as well as electroluminescence (EL) spectra of small area thin film photodiodes and light emitting diodes with size below 1 mm^2. The usefulness of the setup is demonstrated by measurement of optoelectronic properties of a hydrogenated microcrystalline silicon (μc-Si:H) *p-i-n* diode deposited on a semi-transparent nanostructured ZnO:B electrode. No s-shaped *I-V* characteristics were observed under white illumination near an open circuit voltage U_{oc} indicating a negligible charge accumulation near μc-Si:H/ZnO:B interface. The weak infrared EL correlates with the current density.

1. Introduction

Thin film *p-i-n* diodes based on hydrogenated amorphous (a-Si:H) and microcrystalline silicon (μc-Si:H) layers deposited by plasma enhanced chemical vapour deposition (PECVD) are the most common types of thin film diode [1]. The intrinsic layer *i* can be tailored to optimize the depletion region whereas the doped *p*-type and *n*-type layers can be made thin to be opticaly transparent. The thin film diodes have been used for decades in applications such as solar cells, TV screens and scanners [2]. Thin conductive oxides (TCO) are used as optically transparent and electrically conductive electrodes. The nanostructured, boron-doped ZnO layers consisting of pyramidal grains with size about 200 nm combine high optical scattering of visible light, high transparency in the infrared region and moderate sheet resistances of 30 Ω/sq [3, 4].

We have studied extensively nanostructured materials [5]. However, finding technologically compatible materials is challenging [6]. For example, the photoluminescence intensity strongly correlates with the presence of isolated silicon nanoparticles in the mixed amorphous and crystalline phase reaching maximum in samples with crystalline volume fraction at about 50 % and diminishes abruptly in samples with crystalline volume fraction above 80 % prepared by PECVD with SiH_4/H_2 concentration below 5% [7].

The photocurrent spectroscopy reveals that quality of intrinsic layer strongly depends on material type and nanoscale morphology of the substrate [8]. The s-shaped photocurrent is a characteristic signature of a restricted charge transport related to a charge accumulation near one of the electrodes [9]. To study the energy and charge transfer through nanostructured interfaces we need a highly sensitive optoelectronic system for measuring the current-voltage *(I-V)* characteristics as well as photoluminescence (PL) and electroluminescence (EL) spectra in the visible and near infrared region.

In this paper we describe in detail an upgraded setup for a basic optoelectronic characterization of small area photodiodes and demonstrate its usefulness by measuring the *I-V* characteristics and EL spectra of thin film *p-i-n* diodes based on μc-Si:H deposited on semi-transparent nanostructured ZnO:B electrode.

2. Experimental

2.1. Preparation of diode

The diode structure is shown in Figure 1A. A nanostructured ZnO:B thin film, about 2 μm thick was deposited on fused silica glass by a low-pressure metalorganic chemical vapor deposition technique in a single process step by doping with boron only the nucleation stage of the growth [4]. The ZnO:B surface was cleaned 2 min in pure H_2 using 13.56 MHz radio frequency plasma discharge with power 9 W in a stainless steel vacuum deposition chamber with two electrode (capacitive) configuration with a residual pressure of 10^{-5} Pa after 12 hours heating, pumping and degassing [10]. The μc-Si:H layers were deposited from silane SiH_4 (purity 5.0) diborane B_2H_6 and phosphine PH_3 diluted in H_2 (purity 6.0) using 13.56 MHz radio frequency chemical vapour deposition (PECVD), the substrate temperature 220°C, pressure 30 Pa and power 18 W. The H_2 and SiH_4 flow rates were 50 and 2.25 sccm. The 60 nm thick intrinsic μc-Si:H layer was deposited 12 min. The 25 nm thick p-type boron (3 min) and the 25 nm thick n-type phosphorus (5 min) doped layers were prepared using 1% diborane and phosphine addition into the SiH_4. Finally, an array of 12 metallic electrodes (0.8×0.8 mm) was deposited by vacuum evaporation. The selected silane concentration 4.5% corresponds to the hydrogenated silicon layer with high crystallinity and low photoluminescence [7].

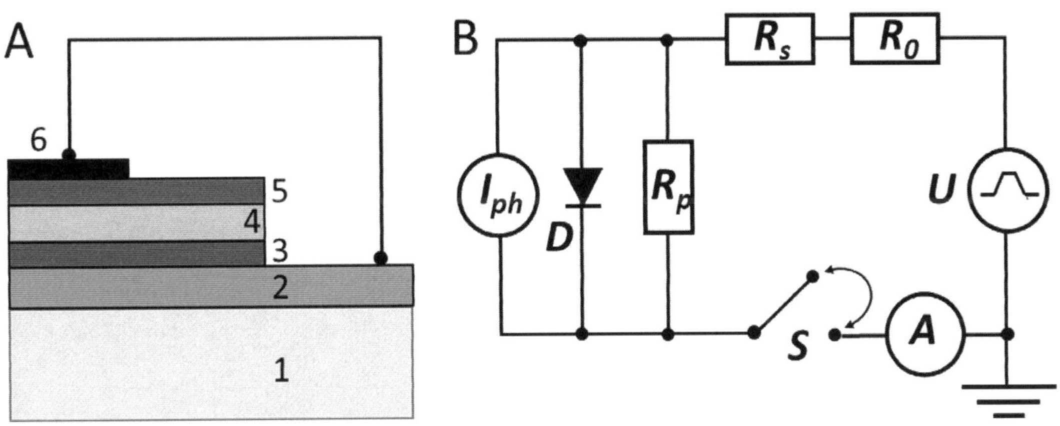

Figure 1. A. A schematic drawing of a thin film p-i-n diode structure. The diode consists of a glass substrate (1), nanostructured ZnO:B (2), *p*-type semiconductor (3), intrinsic semiconductor (4), *n*-type semiconductor (5) and metallic contact (6). **B**. An idealized diode equivalent circuit consist of a parallel resistance R_p, a diode serial resistance R_s and a photocurrent source I_{ph}. The circuit also contains a protective resistor R_0, a switch S, a voltage pulse generator *U* and a picoammeter *A*.

2.2. A setup for current-voltage characteristics

The *I-V* characteristics of diodes were measured in dark and under white light illumination in a pulsed mode using a Keithley #3390 arbitrary waveform generator, a 5½-digit Keithley #6485 picoammeter and a 100 mW/cm^2 white LED synchronized by TTL pulses. Samples were illuminated through glass substrates masked by an aperture to illuminate only a selected diode. The Keithley #3390 arbitrary waveform generator provided arbitrary waveform voltage pulses up to ±10V with the step 20 mV as well as synchronizing TTL pulses. A protective resistor 100 Ω was applied to limit current to 10 mA at

1 V. Unlike digital multimeters, picoammeters offer low input voltage drop (low voltage burden) and selectable input current ranges from 2 nA up to 20 mA. However, due to a relatively high inductance, the picoammeter had to be disconnected from the circuit before changing the current range. This had been done with a PC controlled Signal Recovery Model 3830 multiplexer. After changing the current range, the zero check needed to be performed with the picoammeter input amplifier reconfigured to low input signal. The zero correct feature was enabled to algebraically subtract the voltage offset specific to the selected current range during current measurements.

2.3. A setup for optoelectrical characterization of light emitting diodes

The new setup for optoelectrical characterization of light emitting diodes, shown in Figure 2, evolved from our previous setup designed for steady-state PL spectroscopy in near infrared and visible spectral range 0.75–3 eV [11]. In a new setup, the T-Cube™ LED Driver Thorlabs LEDD1B (max current 1.2 A) drives the Thorlabs MWWHLP1 2W white LED (color temperature 3000 K, full angle 125°). The output power can be varied using the manual control knob and modulated using an external trigger signal. The condenser lens L1 (focal length f=15 mm, diameter Ø 20 mm) concentrates light into parallel beam with adjustable visible light power density up to 100 mW/cm^2. The sample S is illuminated by a parallel beam through an aperture A. The translation XYZ stage (three Thorlabs MT1 Single-Axis Translation Stages) allows to move the sample in all 3 directions with 10 μm precision. The stage, white LED and objective are essential to focus small area diodes onto the input slit of the monochromator. The new setup shares with the previous one the optical detection part including the objective L2 (f = 25 mm, Ø 25 mm, uncoated UV fused silica aspheric Lens, EdmundOptics #48536), monochromator Horiba H20IR with 600 g/mm holographic grating, dichroic mirror DM and photodetectors D1 and D2. The photodetectors have fW sensitivity: Thorlabs PDF10A/M fixed gain amplified Si photodiode D1 with the trans-impedance gain 1pA/V, noise equivalent power NEP 1.4 fW/Hz$^{1/2}$, spectral range 320 – 1100 nm and 0-20 Hz bandwidth and the Thorlabs PDF10C/M fixed gain amplified InGaAs photodiode D2 with the transimpedance gain 10pA/V, NEP 7.5 fW/Hz$^{1/2}$ spectral range 800 – 1700 nm and band width 0–25 Hz. The monochromatic light with the photon energy above 1.3 eV is reflected toward the photodetector D1 via Thorlabs DMLP950R Longpass Dichroic Mirror whereas the monochromatic light with the photon energy below 1.3 eV is transmitted toward the photodetector D2, see Figure 3A. Figure 3A also compares the spectral responsivity of both photodetectors. Both detectors are connected to the Signal Recovery Model 5105 dual phase lock-in amplifier via the Signal Recovery Model 3830 multiplexer that switches between them at the energy 1.3 eV. Since the spectral response was calibrated by the calibrated halogen lamp in wavelength scale, the presentation of spectra in energy scale requires an additional correction [12].

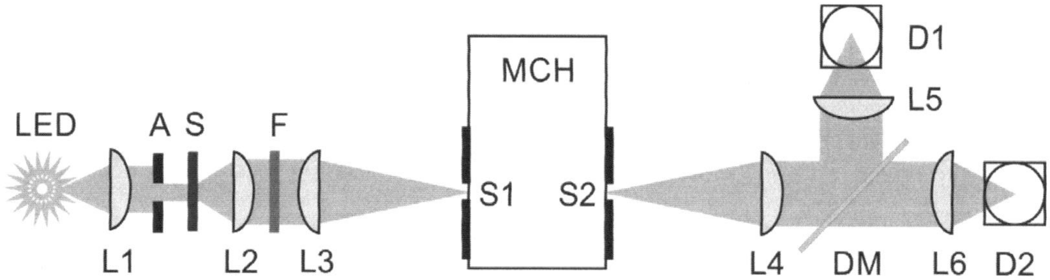

Figure 2. The setup for optoelectrical measurements: Light source LED, condenser lens L1, aperture A, sample S on XYZ translation stage, objective lens L2, optical filter F, focusing lens L3, input slit S1, monochromator MCH, output slit S2, condenser lens L4, dichroic mirror DM, focusing lenses L5 and L6, photodetectors D1 (Si) and D2 (InGaAs). Available slits 0.1, 0.25, 0.5, 1 and 2 mm correspond to the spectral resolution 1, 2, 4, 8, and 16 nm.

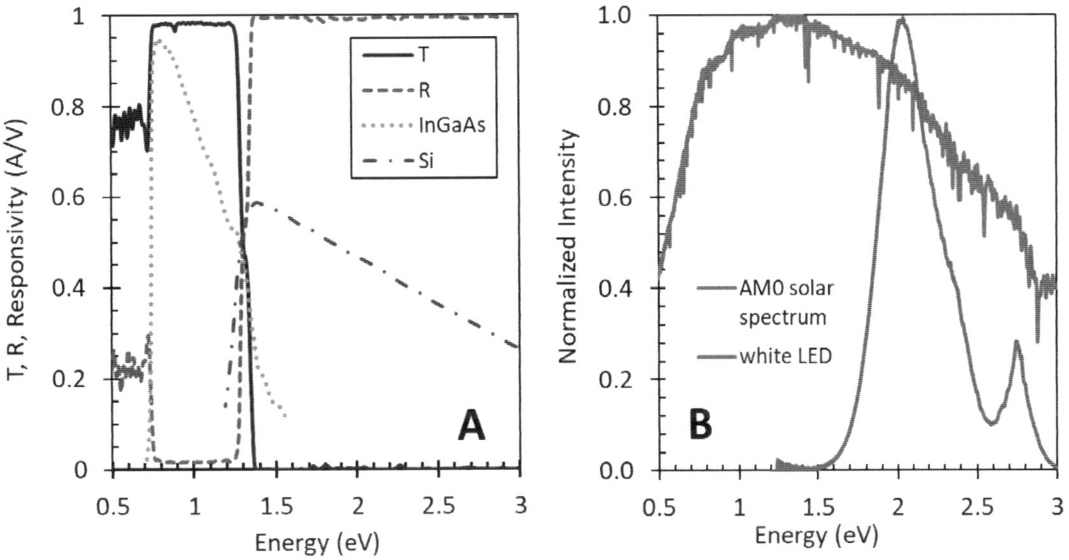

Figure 3. A. Transmittance (T) and reflectance (R) of the dichroic mirror and the spectral responsivity of Si and InGaAs photodiodes. B. The normalised intensity of spectral irradiance of 2W white LED compared to the ASTM E-490 AM0 Standard Solar Spectrum. Data were provided by Thorlabs, Inc. https://www.thorlabs.com/ and https://www.pveducation.org/pvcdrom/appendices/standard-solar-spectra.

Figure 3B compares the normalized intensity of spectral irradiance of the white LED and standard solar spectra. It is obvious that the white LED spectrum differs significantly from the solar spectrum. The white LED produces negligible ultraviolet and infrared light. However, the white LED is useful by providing visible light pulses with adjustable intensity to minimize sample heating.

3. Results and discussion

An idealized diode equivalent circuit is shown in Figure 1B. Its dark *I-V* characteristic is implicitly given by the Equation 1

$$I(U) = I_0\left(e^{\frac{q}{nk_BT}(U-(R_0+R_S)I)} - 1\right) + \frac{U-(R_0+R_S)I}{R_P} - I_{\mathrm{ph}} \qquad (1)$$

Here $q = 1.602\times10^{-19}$ C is the elementary charge, $k_B = 1.381\times10^{-23}$ J/K is the Bolzmann constant, $T = 300$ K the room temperature, n an ideality factor, R_P a parallel resistance, R_S a serial resistance, R_0 a resistance of a protective resistor, U an applied voltage, I a current and I_0 a saturation current. Since the serial resistance of the picoammeter is negligible, the total serial resistance of the circuit is the sum of the resistance R_0 of a protective resistor and a serial resistance of the diode R_S. The ideality factor $n = 1$ is characteristic for diffusion currents whereas the ideality factor $n = 2$ is characteristic for generation-recombination currents [13].

Prior fitting the *I-V* characteristic using Equation 1, a parallel $R_p = \lim\limits_{U\to 0}\left(\frac{dI}{dU}\right)^{-1} = 2\ \mathrm{M\Omega}$ and a total serial resistance $R_S + R_0 = \lim\limits_{U\to\infty}\left(\frac{dI}{dU}\right)^{-1} = 101\ \Omega$ were evaluated from the differential resistance $\left(\frac{dI}{dU}\right)^{-1}$ shown in Figure 5B.

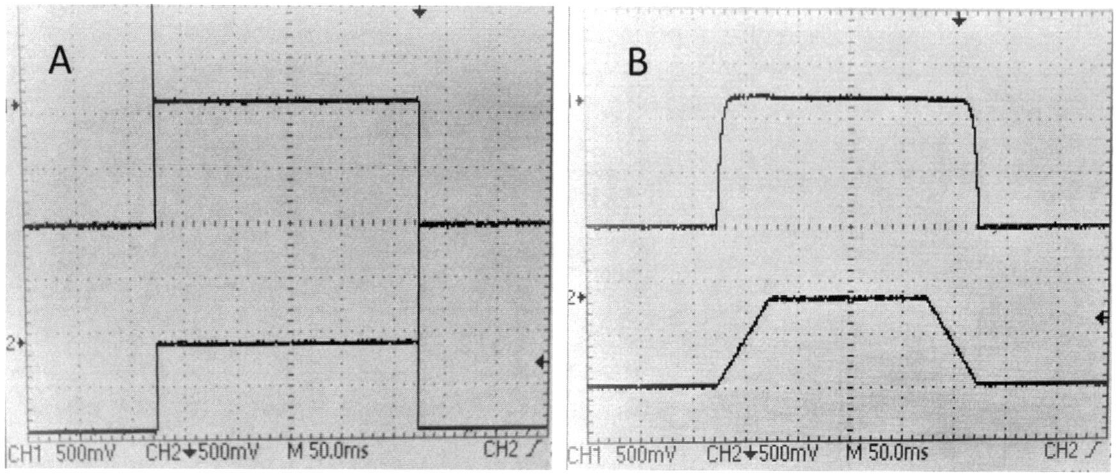

Figure 4. The oscilloscope screen images of rectangular (**A**) and trapezoid (**B**) voltage pulses (CH2, bottom lines) and the corresponding current response (CH1, top). The current response was measured using Tetronix TDS 1002 two channel oscilloscope and the analog output of the picoammeter set to the current range 2μA.

The trapezoid waveform pulses eliminate commutation voltage spikes induced by an inductive load of picoammeter at the rectangular pulse edges, see Figure 4. The optimal frequency for *I-V* measurements was 2 Hz, the picoammeter reading time 100 ms delayed from the trigger pulse by 100 ms.

The non-linear fitting of the dark *I-V* characteristic described by the implicit function $I = I(U, I)$ was carried out for $U \in \langle -0.2, 0.4 \rangle V$ via Wolfram Mathematica software using only two constrained fitting parameters $I_0 > 0$ and $n \in \langle 1,2 \rangle$. The voltage range has been selected for fitting because for small voltages the dark *I-V* characteristics are dominated by leakage and diffusion or generation and recombination currents as described by Equation 1. The fitted results reveal the ideality factor $n \cong 1$, the serial resistance $R_S = 1\ \Omega$ and the saturation current $I_0 = 50$ pA.

The absolute value of the dark current as a function of the applied voltage U is shown in Figure 5A together with fitted curve using Equation 1. The reverse dark currents at $U < -0.2\ V$ are probably dominated by the edge effects due to the small diode area [13]. The $I(U)$ appears to be a linear function for $-0.2\ V < U < 0.2\ V$ indicating that the exponential term is negligible and the diode behaves as an ohmic resistor with resistance R_P and a leakage current $I \approx \frac{U}{R_P}$. Since the saturation current I_0 is much lower then the leakage current, it cannot be directly observed in Figure 5A. The diffusion model fits the measured current well for voltages $0.2\ V < U < 0.4\ V$ whereas the forward current at $U > 0.4\ V$ is dominated by effects not included in Equation 1, such as increase of resistance of doped layers due to high current injection. The forward current is limited at high voltage by the protective resistor R_0.

Figure 6A compares the *I-V* characteristics measured in dark and under visible light illumination in a linear scale. In contrast to our previous paper [14], no S-shape is observed near the open circuit voltage U_{oc} indicating the negligible charge accumulation near μc-Si:H/ZnO interface. Under illumination we detected at zero voltage a photocurrent $I_{ph} = -0.12$ mA corresponding to the current density $J_{ph} = -19$ mA/cm^2, an open circuit voltage $U_{oc} = 0.4$ V, fill factor 61% and an energy conversion efficiency 6%. Thus, our results confirm that nanostructured ZnO thin film has proven to be a good candidate as transparent electrode. It should be noted that the *I-V* characteristics are partly influenced by the presence of the protective resistor R_0. However, due to a relatively low current, the voltage drop at the protective resistor is negligible for the evaluation of I_{ph} and U_{oc}.

The EL spectra and the plot of the total EL integrated in the spectra range 0.75–2 eV as a function of the current density are shown in Figure 6.

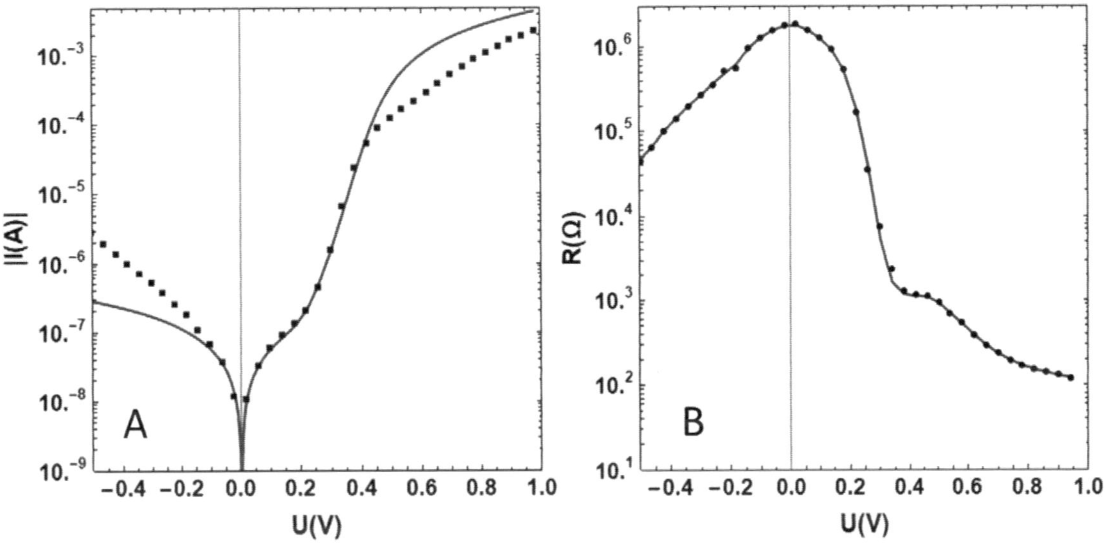

Figure 5. A. The absolute value of a dark current I as a function of applied voltage U (black squares) with the fitted curve (blue solid line). B. The differential resistance $R = \left(\frac{dI}{dU}\right)^{-1}$ (black points) as a function of applied voltage U with the smoothed interpolation curve (blue solid line).

The weak electroluminescence observed in the infrared region diminishes above 1.2 eV. No visible light EL was detected. The inset figure shows that the total intensity of EL increases linearly with forward current density. Thus, the fact that Figure 6 shows the relatively weak EL that is shifted to lower energy is related to high crystallinity of the investigated sample [7].

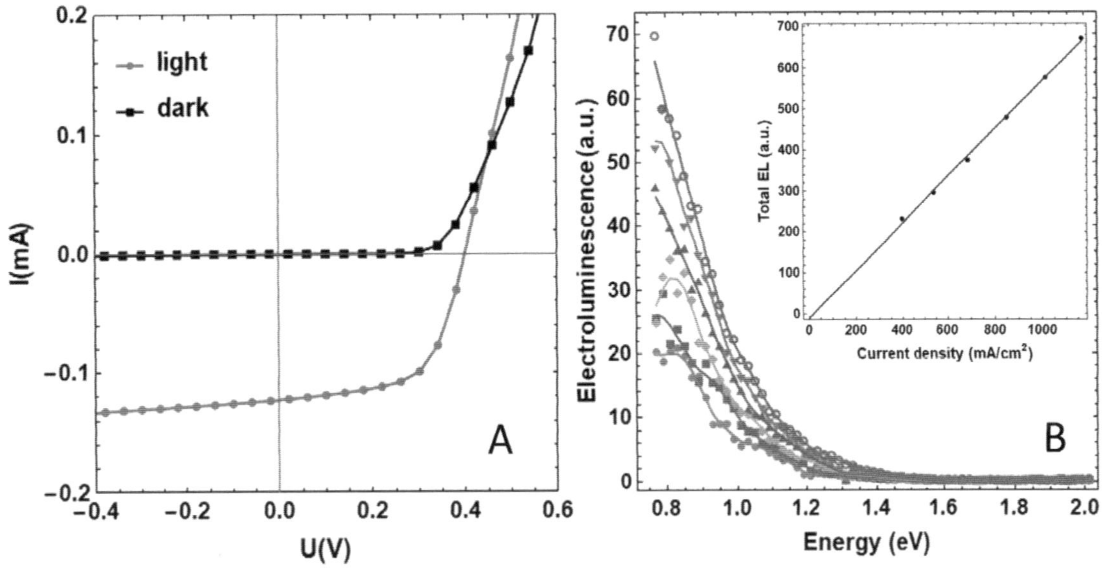

Figure 6. A The I-V characteristics measured in dark (black) and under visible light illumination 100 mW/cm^2 (red). B. The electroluminescence spectra at selected current densities (corresponding to black markers in the inset figure) and the plot of the total electroluminescence in the measured spectra range 0.75–2 eV as a function of the current density together with linear regression model (blue line).

4. Conclusions

Our optoelectronic setup previously developed for photoluminescence spectroscopy in the visible and near infrared region has been extended for measurements of optoelectronic properties of small area thin film photodiodes using an arbitrary waveform generator as a pulse voltage source, a white light source and a picoammeter operating in triggered pulse mode. Three preventive measures were successfully implemented to avoid diode damage: a protective resistor to limit the maximum current; trapezoidal voltage pulses to avoid commutation spikes and a switch disconnecting the picoammeter during the current range change to isolate its high inductance load. Using the setup, we have observed that the dark *I-V* characteristics of thin film *p-i-n* diodes based on μc-Si:H with high crystallinity content deposited on semi-transparent nanostructured ZnO:B electrode are dominated by leakage and diffusion currents for small voltages. No s-shaped photocurrent has been observed near the open circuit voltage U_{oc} indicating the negligible charge accumulation near μc-Si:H/ZnO:B interface. The weak EL appears only in the infrared region and diminishes above 1.2 eV. The low EL background makes the diode suitable for future studies of a radiative recombination related to a charge transfer at an interface between nanostructured surface and a surface of embedded nanoparticles.

Acknowledgement

The work was supported by the Czech Science Foundation project 19-02858J and by the Operational Programme Research, Development and Education financed by European Structural and Investment Funds and the Czech Ministry of Education, Youth and Sports (Project No. SOLID21 - CZ.02.1.01/0.0/0.0/16_019/0000760). We acknowledge M. Benkhaira, a technician from EPFL-IMT-PVlab Photovoltaic and Thin-Film Electronics Laboratory, Rue De La Maladière 71b, 2002 Neuchâtel, Switzerland who prepared nanostructured ZnO:B substrates.

References

[1] Smets A H M, Jäger K, Isabella O, Swaaij R A C M M van and Zeman M 2016 *Solar energy: the physics and engineering of photovoltaic conversion, technologies and systems* (Cambridge, England: UIT Cambridge Ltd.)

[2] Street R A 1991 *Hydrogenated amorphous silicon* (New York: Cambridge University Press)

[3] Faÿ S, Steinhauser J, Oliveira N, Vallat-Sauvain E and Ballif C 2007 Opto-electronic properties of rough LP-CVD ZnO:B for use as TCO in thin-film silicon solar cells *Thin Solid Films* **515** 8558–61

[4] Ding L, Boccard M, Bugnon G, Benkhaira M, Nicolay S, Despeisse M, Meillaud F and Ballif C 2012 Highly transparent ZnO bilayers by LP-MOCVD as front electrodes for thin-film micromorph silicon solar cells *Sol. Energy Mater. Sol. Cells* **98** 331–6

[5] Mičová J, Buryi M, Šimek D, Drahokoupil J, Neykova N, Chang Y-Y, Remeš Z, Pop-Georgievski O, Svoboda J and Im C 2018 Synthesis of zinc oxide nanostructures and comparison of their crystal quality *Appl. Surf. Sci.* **461** 190-5

[6] Ventosinos F, Klusacek J, Finsterle T, Kunzel K, Haug F-J and Holovsky J 2018 Shunt Quenching and Concept of Independent Global Shunt in Multijunction Solar Cells *IEEE J. Photovolt.* **8** 1005–10

[7] Remes Z, Stuchlik J, Purkrt A, Ledinsky M and Kupcik J 2017 Near infrared photoluminescence of the hydrogenated amorphous silicon thin films with in-situ embedded silicon nanoparticles *Ceram. - Silik.* **61** 136–40

[8] Holovský J, Dagkaldiran Ü, Remeš Z, Purkrt A, Ižák T, Poruba A and Vaněček M 2010 Fourier transform photocurrent measurement of thin silicon films on rough, conductive and opaque substrates *Phys. Status Solidi A* **207** 578–81

[9] Wagenpfahl A, Rauh D, Binder M, Deibel C and Dyakonov V 2010 S-shaped current-voltage characteristics of organic solar devices *Phys. Rev. B* **82** 115306

[10] Stuchlik J, Stuchlikova T H, Remes Z, Purkrt A, Fajgar R, Kostejn M, Kupcik J, Zhuravlev K, Sveshnikova L, Galkin K N and Galkin N G 2015 The Combined Technological Methods for Deposition of Si:H Thin Films and Structures with Embedded NPs *Adv. Sci. Eng. Med.* **7** 265–9

[11] Remes Z, Micova J, Krist P, Chvatil D, Effenberg R and Nesladek M 2015 N-V-related fluorescence of the monoenergetic high-energy electron-irradiated diamond nanoparticles: N-V-related fluorescence of electron-irradiated diamond nanoparticles *Phys. Status Solidi A* **212** 2519–2524

[12] Mooney J and Kambhampati P 2013 Get the Basics Right: Jacobian Conversion of Wavelength and Energy Scales for Quantitative Analysis of Emission Spectra *J. Phys. Chem. Lett.* **4** 3316–8

[13] Sze S M and Ng K K 2007 *Physics of semiconductor devices* (Hoboken, N.J: Wiley-Interscience)

[14] Remes Z, Stuchlik J, Stuchlikova H T, Dragounova K, Ashcheulov P, Taylor A, Mortet V and Poruba A 2019 Optoelectronic Properties of Hydrogenated Amorphous Substoichiometric Silicon Carbide with Low Carbon Content Deposited on semi-Transparent Boron-Doped Diamond *Phys. Status Solidi A* pssa.201900241

The effect of temperature on bacterial self-healing processes in building materials

P Ryparová, P Tesárek, H Schreiberová and Z Prošek

Czech Technical University in Prague, Faculty of Civil Engineering, Thákurova 7, Prague 6, Czech Republic

pavla.ryparova@fsv.cvut.cz

Abstract. This paper is focused on the bacterial induced calcitation for the crack healing. The bacteria applied for this purpose are from group which is adapted for growth in the high pH environment as a concrete in hydration phase and their metabolic activity leads to create of calcite. In this study, three different bacteria strains (*Sporosarcina pasteurii, Bacillus cohnii, Bacillus pseudofirmus)* were applied and the influence of various temperatures on their microbial properties was investigated. Our previous experiment indicated that one of the applied bacterial strain in spores form (*Bacillus pseudofirmus*) are able to survive the temperatures in the range from −20 °C to 140 °C. The experiment described in this paper extends the previous study and determines the effect of different temperatures on the change in growth activity. In this experiment, bacterial activity was determined based on the change of absorbance in 640 nm by spectrophotometric measurements. The experiment was performed at optimal temperature (30 °C) and lower temperature (10 °C) and it used suitable broth for calcitation. The results showed that the beginning of metabolic activity was shifted by 40 to 50 hours. Only *Bacillus cohnii* showed different results because its metabolic activity was nearly zero at 10 °C.

1. Introduction

Concrete is the most used construction material worldwide. Degradation of this material is known and described phenomena. The main degradation processes are caused by pollutants present in the atmosphere such as oxides of nitrogen, carbon, sulphur, acids and many other compounds. These processes are known as atmospheric corrosion or chemical degradation. Nevertheless, even though concrete structures generally undergo a number of degradation processes, they are designed to maintain their service life for more than 50 years [1].

These mentioned problems lead to higher porosity and subsequent to increase concentration of aggressive substances in pore system. The aforementioned compounds mainly react with free calcium ions which results in a decrease in pH, thus durability of the material is endangered. Nowadays, many types of repair systems are available. However, these epoxy-based fillers or silane-based water repellent systems may have only short-term effectiveness or a negative impact on the environment. Today it focuses more on instant repair or self-healing of cracks may extend the service-life of concrete structures [2].

Bacteria can be found everywhere and some of them are resistant enough to survive in various harsh conditions. The self-healing of concrete can be caused by some autogenic processes or induce by bacteria. In both cases, the main reactions lead to creation of new crystals of calcium carbonate

($CaCO_3$). Microbially induced calcification (MIC) is a novel innovative alternative method for protection of concrete construction against change of porosity. This phenomenon has been known for many years [3-5]. Ramakrishnan published a pilot study which focused on bioremediation as a treatment of cracked concrete [6]. Some bacterial strains can induce a reaction between atmospheric carbon dioxide and calcium ions, see Equation 1. Other ones can use the hydrolysis of urea which is catalysed by urease. As a consequence, urea is degraded into carbonate and ammonium, resulting in an increase in the pH and carbonate concentration in the bacterial environment [7], see Equation. 2.

$$H_2O + CO_2 \rightarrow HCO_3^- + H^+$$
$$Ca^{2+} + 2HCO_3^- \rightarrow CaCO_3 + H^+ + HCO_3^- \rightarrow CaCO_3 + H_2O + CO_2 \quad (1)$$

or with urea

$$CO(NH_2)_2 + 2H_2O \rightarrow 2NH_4^+ + CO_3^{2-}$$
$$CO_3^{2-} + Ca^{2+} \leftrightarrow CaCO_3 \quad (2)$$

In this paper, the bacteria for microbially induced calcification were selected from groups which are adapted to colonization in a high pH environment and can produce calcium carbonate (*Sporosarcina pasteurii, Bacillus cohnii, Bacillus pseudofirmus). Sporosarcina pasteurii* is a gram-positive bacterium able to survive in highly alkaline environments (pH ~ 10) and has ability to secrete a copious amount of enzyme urease and use the second aforementioned equilibrium (Eq. 2). The group *Bacillus* includes facultative and obligate alkaliphilic species that grow well at pH values (higher than 9.0). Alkaliphilic *Bacillus* strains secrete large amounts of alkaline proteases with a high stability at elevated pH and temperature values.

This novel strategy for preventing crack formation seems to be the right direction to minimize repair costs of constructions. The bacteria which we mentioned above have great potential for self-healing due to their beneficial characteristics (pH resistance and ability to produce calcite). However, previous studies pointed out several problematic aspects which need to be solved e.g. the application of bacteria with non-reduction of material properties and bacteria survival in the first period of cement hydration. Further, the investigations must focus on the restart of the bacterial life cycle and determine the necessary conditions for successful bacterially induced calcification. The main question is the minimum time for water to be in pore system of material in specific conditions (T, pH, strain bacteria, source of calcium). This article follows the previous experiment which dealt with the survival of bacteria after freeze-thaw cycles between -20 °C to 5 °C [8]. The current study wants to determinate the necessary time to restart the bacterial life cycle in temperatures between 10 °C and 30 °C.

2. Materials and methods

The selected bacteria were obtained from Czech collection of microorganisms (CCM) or from Belgian Coordinated Collections of Micro-organisms (BCCM). The access number of strains were: *Sporosarcina pasteurii CCM 2056, Bacillus cohnii CCM 4369, Bacillus pseudofirmus* (LMG 17944). The used culture media are described in Table 1. The media were prepared based on the prescriptions from CCM or BCCM.The medium 53 contain following compounds: Lab-Lemco' beef extract 1 g/l, yeast extract 2 g/l, peptone 5 g/l, NaCl 5 g/l, agar 15 g/l and the pH 9.7 was achieved by Na-sesquicarbonate solution after sterilization in an autoclave.

Table 1. The correlation between bacterial strain, media and optimal temperature.

Strain	Optimal temperature [°C]	Growth medium
Bacillus cohnii	37	53
Sporosarcina pasteurii	30	B13
Bacillus pseudofirmus	30	253

The Na-sesquicarbonate solution consisted of NaHCO3 4.2 g, Na_2CO_3 anhydrous 5.3 g in distilled water 100 ml, the sterilization was performed by filtration. The medium 253 was prepared based on the medium 53 only with 0.5% NaCl. The medium B13 contained following compounds: Lab-Lemco' beef extract 1 g/l, yeast extract 2 g/l, peptone 5 g, NaCl 5 g/l, agar 15 g/l, urea 20 g/l. The pH value was adjusted to 7.4 after sterilization by Na-sesquicarbonate solution described above.

The bacterial inoculation for experiments were prepared by three time repeated follow cultivation for starting the maximal growth rate in optimal condition for seven days. Further, this inoculum was applied in a ratio 1 ml of the inoculum to 100 ml of the fresh media for the biological oxygen demand (BOD) experiments and 1 ml of the inoculum to 50 ml of the fresh media for the ELISA experiments.

2.1. ELISA measurement

The amount of bacteria cells or multiplication were measured by change absorbance in 630 nm. The values were recorded every hour in the daytime for the first 5 days, then the data collection frequency decreased to 5 times in the daytime. This obtained data was used for determinate of growth curve. The whole experiment was performed for 14 days.

2.2. BOD experiments

The Biological Oxygen Demand (BOD) method is based on the measurements of the amount of oxygen by consumed by aerobic microorganisms. In our study, this method was used for the determination of metabolic activity of the investigated bacteria. The BOD values are most commonly expressed in milligrams of oxygen consumed per litter of a sample during 5 days of incubation at 20 °C. The BOD device is standardly calibrated to temperature around 25 °C. For this reason, the BOD method must have been firstly tested on a known bacterial strain in other than the calibration temperature in order to determine its applicability in our experiment. The bottles for BOD analysis were filled with 100 ml of fresh media respectively tempered to the selected temperature. Subsequently, the bacterial inoculums in a volume of 2 ml were added respectively. The bottles were closed and placed into the cultivation temperature. The measurements were continuous (i. e. once every 0.5 h for 5 days).

3. Results

The influence of different temperatures (10 °C and 30 °C) on the bacterial growth was determined as a change of optical density by ELISA. The bacteria *Bacillus pseudofirmus* (Figure 1) had a longer log phase at the lower temperature but the optical density was similar in 145 hours of experiment.

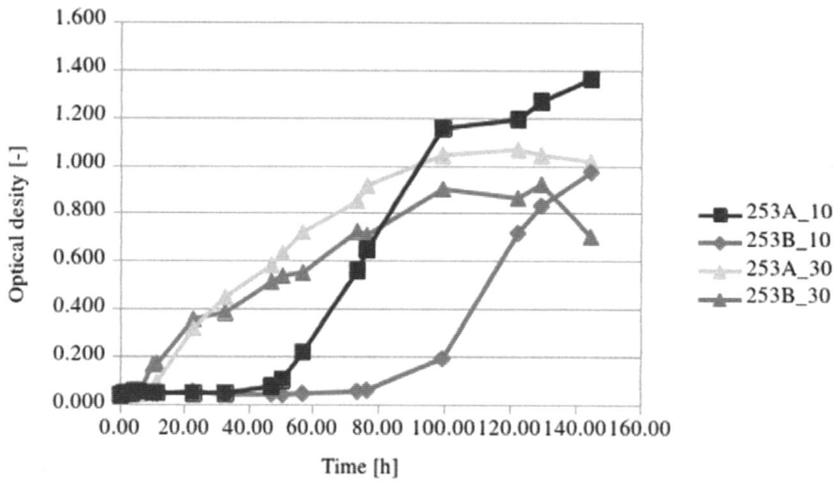

Figure 1. The growth curve for *Bacillus pseudofirmus*, the number after label is the used temperature for experiment, 10 °C or 30 °C.

The ending of the log phase of the *Bacillus pseudofirmus* cultivation at 10 °C was between 45 – 76.5 hours. The log phase of the *Bacillus pseudofirmus* cultivation at 30 °C was shortened to 10.5 hours.

For the results for ELISA test of bacteria *Sporosarcina pasteuri* (Figure 2). The log phase was shifted as a results for *Bacillus pseudofirmus.* The bacteria growth has started in range 32.5 to 50.6 hours in 10°C and in 22.5 hours in the 30 °C experiment. The limited supply of nutrients was detected in 140 hours. The metabolic activity measured by BOD started during the 1st hour for the 10 °C cultivation and during 30 minutes for the 30 °C cultivation. This activity was stable throughout the whole experiment.

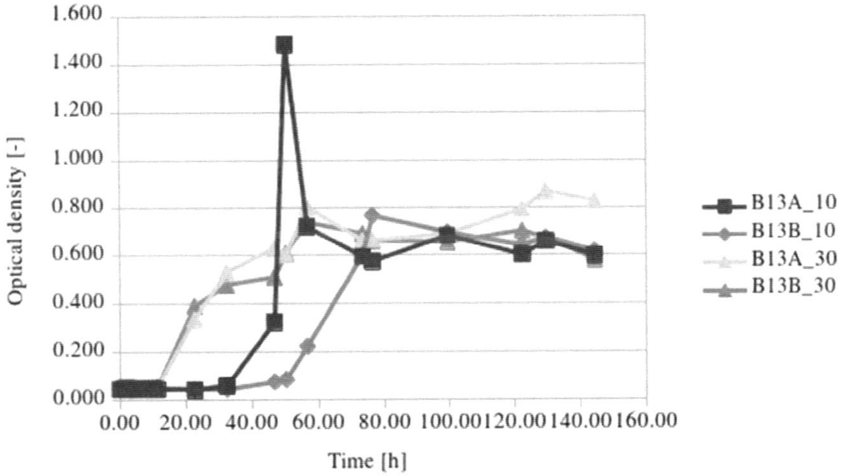

Figure 2. The growth curve for Sporosarcina pasteurii, the number after label is the used temperature for experiment, 10 °C or 30 °C.

The multiplication of *Bacillus cohnii* can be seen in Figure 3. This cultivation has different results from the other investigated bacteria. The log phase was the longest of all used bacterial strain The bacteria at 10 °C did not start the growth until 150 hours.. The cultivation at 30 °C showed very different results in comparison with the 10 °C results. The start of multiplication was between 32.5 and 122.5 hours.

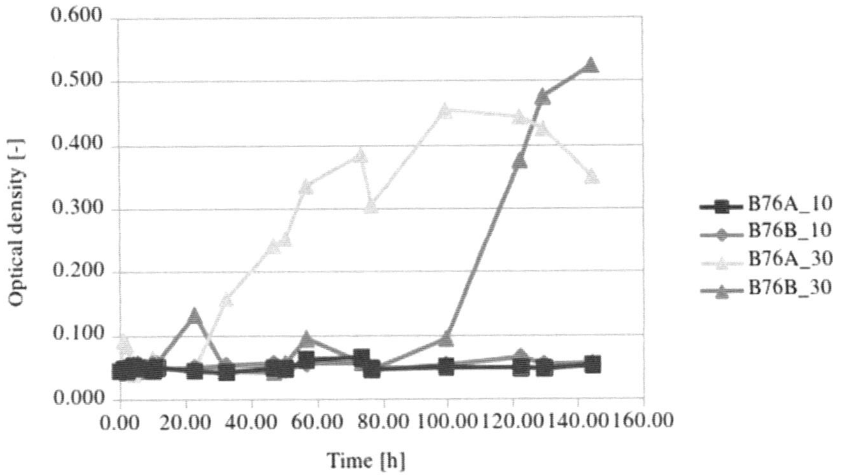

Figure 3. The growth curve for *Bacillus cohni*, the number after label is the used temperature for experiment, 10 °C or 30 °C

The metabolic activity measured by BOD for this temperature was lower but stable in the whole experiment. The metabolic activity measured by BOD system showed lower value, but they were stable for 140 hours. These results show that *Bacillus cohnii* have not stable growth characteristics and also our previous unpublished measurements indicated that the bacteria have very unstable growing activity in pH more than 8.5.

4. Conclusion

In this study, three strains of bacteria were studied and the possibility of their application for the bacterially induced calcification was determined. All of the selected bacteria are from the class *Bacilli*, two of them are from genus *Bacillus* and one from *Sporosarcina*. These bacteria are isolated from environment with high alkalinity therefore they have properties to survive during hydration of concrete. These three strains can survive at temperature under -5°C in the form of spores, and they further restart their metabolic activity when the conditions become optimal. The time required to the metabolic activity recovery depends on the water content, nutrient source and temperature. As it has been expected, the results of metabolic activity in lower temperatures (around 10 °C) showed that the recovery time is longer as it between 32.5 to 76.5 hours for *Sporosarcina pasteurii* and *Bacillus pseudofirmus*. The strain *Bacillus cohnii* did not recover multiplication in our experiment. The higher temperature is approaching their optimal temperature and their recovery was faster. It is around 20 hours. The strain *Bacillus cohnii* had longer period in one of the cases but the results indicated the strains instability in the high alkalinity conditions.In this study, the results showed that temperature around 10 °C is sufficient for the recovery of metabolic activity of two of the applied bacterial strains (*Sporosarcina pasteurii* and *Bacillus pseudofirmus*). but the time require for wetness material are longer. The temperatures closer to optimal showed better results as expected. However, maintaining high water content at this temperature in real environment is very problematic. The further studies should deal with protection of bacteria during hydration period and keep.

Acknowledgement
This research is based upon work supported by the Czech Grant Foundation under Grant Number GA CR 18-15697S.

References

[1] Emmons P H and Sordyl D J 2006 The state of the concrete repair industry and a vision for its future *Concrete Repair Bulletin* 7–14
[2] Wang J Y, Sones H, Verstraete W and De Belie N. 2014 Self-healing concrete by use of microencapsulated bacterial spores *Cement Concrete Res.* **56** 139–152
[3] Edvardsen C 1999 Water permeability and autogenous healing of cracks in concrete *ACI Mater. J.* **96** 448–454
[4] Huang H L, Ye G and Damidot D 2013 Characterization and quantification of self-healing behaviors of microcracks due to further hydration in cement paste. *Cement Concrete Res.* **52** 71–81
[5] Jacobsen S, Sellevold E J 1996 Self healing of high strength concrete after deterioration by freeze/thaw. *Cement Concrete Res.* **26** 55–62
[6] Ramakrishnan V, Bang S S and Deo K S 1998 A novel technique for repairing cracks in high performance concrete using bacteria, in Proceeding of International Conference on High Performance High Strength Concrete (Perth,WA) 597–618
[7] Fischer S, Galinat J K, Bang S S 1999 Microbiological precipitation of $CaCO_3$ *Soil Biol. Biochem.* **31** (11) 1563-71
[8] Schreiberová H, Bílý P and Ryparová P. 2018 Applicability of bio-based selfhealing concrete in Central European conditions: A preliminary study, in Proceeding of RILEM TC 253-MCI International.

IOP Conf. Series: Materials Science and Engineering **726** (2020) 012013 doi:10.1088/1757-899X/726/1/012013

Deposition and characterization of Ti-Al-C-N coatings

M Sahul[1], M Haršáni[2], P Babincová[1], Ľ Čaplovič[1], M Sahul[3] and P Drobný[1]

[1]Slovak University of Technology in Bratislava, Faculty of Materials Science and Technology in Trnava, Institute of Materials Science, Ulica Jána Bottu č. 2781/25, 917 24 Trnava, Slovak Republic
[2]Staton, s.r.o., Sadová 1148, 038 53 Turany, Slovakia
[3]Slovak University of Technology in Bratislava, Faculty of Materials Science and Technology in Trnava, Institute of Production Technologies, Ulica Jána Bottu č. 2781/ 25, 917 24 Trnava, Slovak Republic

martin.sahul@stuba.sk

Abstract. In the present work, Ti-Al-C-N coatings were deposited on cemented carbide substrates by lateral rotating cathodes (LARC®) process using Platit π^{80}+DLC deposition unit. The effect of C_2H_2 gas flow rate on elemental and phase composition, deposition rate, cross-sectional and surface morphology, mechanical and tribological properties of the coatings was studied. Following analytical techniques, namely: scanning electron microscopy (SEM) with energy and wave dispersive X-ray spectroscopy (EDS and WDS), X-ray diffraction analysis (XRD), nanoindentation measurements, Rockwell C indentation test and tribological testing were used for Ti-Al-C-N coatings evaluation. From the EDS analysis, it was found that the carbon content in the coatings increased from 0 at.% to 22.3 at.% as the C_2H_2 gas flow rate increased from 0 sccm to 75 sccm. The increase in deposition rate of coatings from 0.029 μm/min to 0.052 μm/min was documented. From XRD results it was found that the coatings consist of a cubic B1-NaCl type $Ti_1N_{0.9}$ phase. The maximum hardness was observed at C_2H_2 gas flow rate of 25 sccm and the lowest friction coefficient (0.35) at the maximum C_2H_2 gas flow rate. The coatings deposited at C_2H_2 gas flow rates (25 sccm and 50 sccm) exhibited an excellent adhesion.

1. Introduction
Currently, hard coatings fabricated by methods of physical vapor deposition (PVD) have been strongly used to improve properties of various kinds of engineering materials and are now widely employed in numerous technical fields [1-4]. Hard coatings are characterized by excellent mechanical properties, such as high hardness, wear resistance and thermal stability [5-8]. Titanium Nitride (TiN) coatings belong to the group of the most common used PVD hard coatings today [9,10]. However, one of the major drawbacks of TiN coatings is their limited oxidation resistance (up to 500 °C) [3]. This problem has been solved by incorporation of aluminium in the cubic NaCl-type TiN matrix, which leads to creation of the Ti-Al-N metastable solid solution [11-15]. Depending on the Al content, the supersaturated Ti-Al-N phase decompose (~800 °C) via spinodal decomposition into two phases: fcc-TiN and fcc-AlN [16-18]. In addition, some researchers has shown an enhancement of tribological and mechanical properties by the addition of carbon to Ti-Al-N coatings [19-21]. Stueber et al. deposited using reactive magnetron sputtering series of Ti-Al-C-N coatings with different carbon contents (0-28 at.%). It resulted to deposition of nanocomposite coatings with a structure of a metastable hard

nanocrystalline fcc-TiAlCN phase embedded in an amorphous carbon matrix (a-C) [22]. In the work of Lackner et al., Ti-Al-N and Ti-Al-C-N coatings were fabricated by employing the pulsed laser deposition (PLD) technique. They found that the variation of the deposition parameters had caused a change of the chemical composition, the texture and the crystallinity of the coatings and, consequently, the mechanical properties and tribological behaviour, which was characterized by pin-on-disc test. The results proved the excellent industrial applicability of these coatings for cold-forming operations. Very low-wear rates were found for the Ti-Al-N coatings. In contrast, the Ti-Al-C-N coatings possessed low-friction coefficients of approximately 0.2 [21]. The effect of magnetic filtering on morphology, surface roughness, chemical composition, adhesion and corrosion resistance of Ti-Al-C-N coatings during a filtered cathodic arc deposition (FCAD) was investigated by Hsu et. al. Despite of the deposition rate of the filtered Ti-Al-C-N coating was slower compared to the unfiltered one, the former had a relatively denser structure, smoother surface and exhibited better adhesion. In addition, both coatings improved the corrosion resistance of DC53 steel in 3.5 wt.% NaCl solution. Particularly, filtered Ti-Al-C-N coating offered more effective corrosion protection than unfiltered Ti-Al-C-N [23]. Al-Bukhaiti et al. used the magnetron sputtering for a deposition of Ti-Al-C-N coatings in the form of multilayers consisted of Ti/Ti-Al-N/Ti-Al-C-N nanoscale layers. The tribological performance of the coatings was studied by a ball-on-disk test against 100Cr6 steel and Al_2O_3 balls. Results showed that the deposited coatings were characterized by low wear rates of 10^{-15} m^3/Nm, low friction coefficients in the range of 0.25-0.37 and satisfactory hardness in the range of 17-20 GPa [24].

In this study, Ti-Al-C-N coatings with different C_2H_2 gas flow rates were deposited by lateral rotating cathodes method (LARC®). The influence of C_2H_2 gas flow rate on microstructure, morphology, chemical and phase composition, nanoindentation hardness, adhesion and tribological properties of Ti-Al-C-N coatings was studied.

2. Experimental details

2.1. Deposition

PLATIT π^{80}+DLC deposition unit that uses technology of lateral rotating cathodes arc (LARC®) was utilized for depositions of Ti-Al-C-N coatings. The LARC® technique has two major advantages over the conventional planar vacuum arc deposition systems. The first advantage comes from the unique design and positioning of cylindrical cathodes, which results in a maximum effective cathode surface and longer cathode lifetime. The second advantage comes from the Virtual Shutter function, which makes it possible for in situ cathode cleaning before coating deposition without a plasma interruption. This two features result in superior adhesion and smooth, droplet free surfaces of the as-deposited coatings [25]. Mirror polished tungsten carbide disks (WC-10 wt.% Co) of 12 mm in diameter were used as substrates. Prior to the depositions, the substrates were ultrasonically cleaned in acetone, dryed in a hot air and mounted onto a rotational substrate holder. The deposition chamber was pumped down to a pressure of 10^{-2} Pa by using rotary and turbomolecular pumps and heated up to the temperature of 470°C. The substrates were cleaned by Ar^+ glow discharge and then, a thin TiN layer was deposited in order to improve the coatings adhesion before the deposition. The constant deposition parameters were following: Ti cathode current 132 A, Al cathode current 95 A, substrate bias −90 V, substrate holder rotation 12 rpm and N_2 gas flow rate 100 sccm. C_2H_2 gas flow rate was varied from 0 to 75 sccm in order to adjust the carbon content in the coatings. The influence of C_2H_2 gas flow rate and thus carbon content in the coatings on their chemical and phase composition and mechanical and tribological properties was evaluated. Deposition time (90 min.) was the same for all deposition cycles.

2.2. Characterization

The elemental composition of the Ti-Al-C-N coatings was simultaneously measured with INCA X-Max 50 mm^2 energy dispersive X-ray spectrometer to determine heavier elements (Ti and Al) and INCA Wave wave energy dispersive spectrometer (light elements C and N), both manufactured by Oxford Instruments Company. The fracture areas, surface morphologies and indents after Rockwell C indentation test were observed by JEOL JSM 7600F high resolution scanning electron microscope in secondary electrons imaging mode. Parameters during observation were following: $U = 20$ keV, $I = 2$ nA and WD = 15 mm. Identification of phases present in the deposited coatings was performed by X-ray diffraction analysis using an Empyrean diffractometer from Panalytical Company with a CoK_α anode ($\lambda = 0.178897$ nm). The analysis was carried out in Bragg-Brentano geometry. The parameters for the Co anode used during the measurement were following: $U = 40$ kV, $I = 40$ mA, angular interval 20-130° 2θ and step size 0.026° 2θ. Nanoindentation measurements were performed using Anton Paar NHT2 device with Berkovich diamond indenter. The load was gradually increased to 20 mN at a loading rate of 60 mN/min and was held at the maximum value for 5 s. Continuous stiffness method was employed to assure that the indentation depth was within the 5-10% of the coating thickness. The indentation curves were processed with Oliver & Pharr method. The average friction coefficient was measured using ball-on-disc sliding test with a Bruker UMT TriboLab device. Tungsten carbide balls with diameter of 6.3 mm were used as a counterpart material. The sliding tests were performed with a sliding speed of 0.15 m/s under a load of 5 N. For purposes of the Rockwell C indentation test, a common hardness tester produced by Škoda Company was utilized.

3. Results and discussion

3.1. Elemental composition

Elemental composition of the Ti-Al-C-N coatings deposited with increasing C_2H_2 gas flow rate is documented in Figure 1. Not surprisingly, as C_2H_2 gas flow increased, the C content in the coatings increases from 0 at.% to 18.72 at.%, while N content decreases from 54.28 at.% to 47.34 at.%. Meanwhile, the Ti content also decreases correspondingly from 22.89 at.% to 12.35 at.%. The Al content remains basically unchanged (~21 at.%).

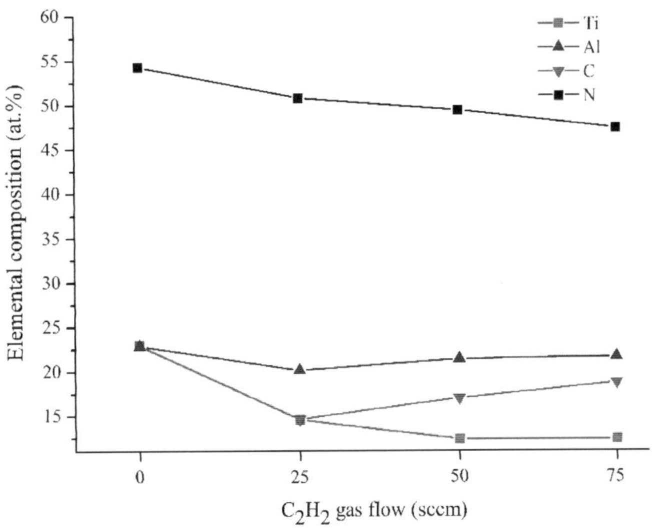

Figure 1. Elemental composition of Ti-Al-C-N coatings as a function of C_2H_2 gas flow rate in at.%.

3.2. Cross sections and surface morphology

The morphology of the cross sections of Ti-Al-C-N coatings fabricated at different C_2H_2 gas flow rate is given in Figure 2. A thin TiN adhesive layer was deposited to the substrate in order to improve the adhesion of the coatings. Each of Ti-Al-C-N coatings are characterized by the smooth surface topography and columnar structure with the columns diameter widens as the coatings grow. Higher deposition rate (from 0.029 μm/min to 0.052 μm/min) was documented with increasing of C_2H_2 gas flow rate. The maximum coating thickness of ~4.71 μm was observed at C_2H_2 gas flow of 75 sccm (Figure 2d). The presence of micro-droplets that are associated with the LARC® process was documented as well (Figure 3). Micro-droplets are being formed during the deposition process due to the ejection of liquid metal particles from the cathodes and may negative affect the coating during its growth [26].

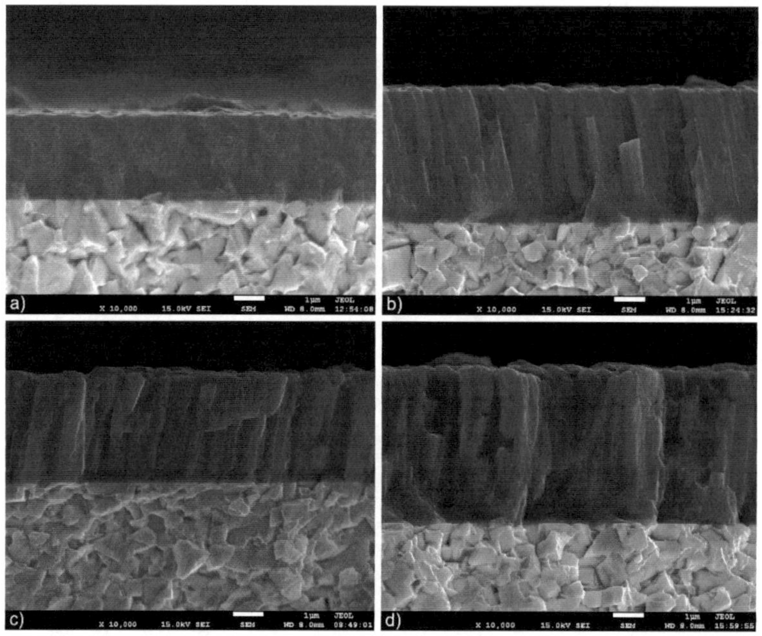

Figure 2. Cross-sections of Ti-Al-C-N coatings deposited at different C_2H_2 gas flow rate, a) 0 sccm, b) 25 sccm, c) 50 sccm and d) 75 sccm.

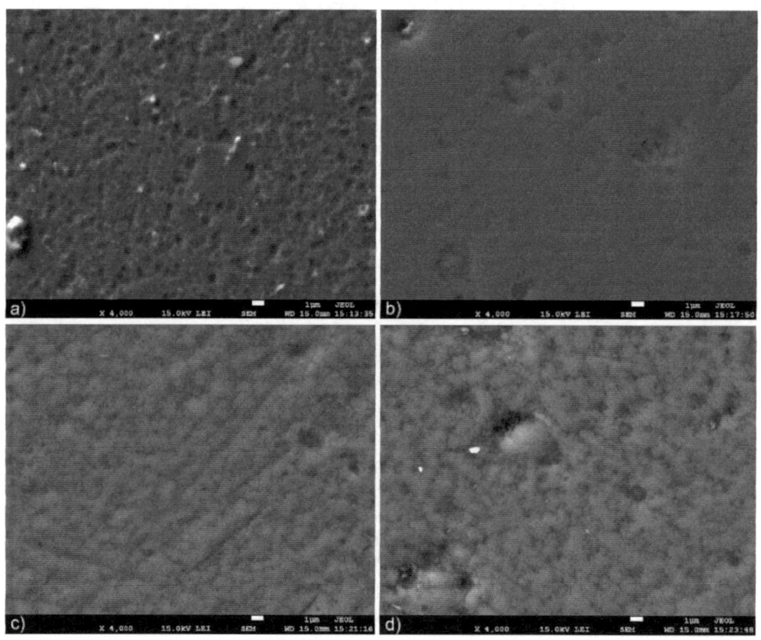

Figure 3. Surface morphology of Ti-Al-C-N coatings deposited at different C_2H_2 gas flow rate, a) 0 sccm, b) 25 sccm, c) 50 sccm and d) 75 sccm.

3.3. Structure

XRD patterns of Ti-Al-C-N coatings deposited at different C_2H_2 gas flow rates are given in Figure 4. Identification of peaks confirmed the assumption that the Ti-Al-C-N coatings consist of a cubic B1-NaCl type $Ti_1N_{0.9}$ phase (ICSD 98-000-1547) with a maximum intensity of (111) diffracting plane at a Bragg angle of 42.876° 2Θ. The unit cell parameters are following $a = b = c = 4.239$ Å, $\alpha = \beta = \gamma = 90°$. The carbon free coating is strongly oriented in the [111] direction. The increase in the C_2H_2 gas flow rate caused a substitution of N atoms by C atoms and decrease in intensity of (111) $Ti_1N_{0.95}$ and (022) $Ti_1N_{0.95}$ diffracting planes, because C could act as a diffusion inhibitor that restrains the growth of the coating grains, what can result in presence of randomly oriented grains in the coatings [27]. Because of the atomic volume of C is larger compared to that of N, the Ti-Al-N unit cell increased with increasing of the C_2H_2 gas flow rate. In addition, the peaks position shift to a lower diffraction angles is caused by increase of carbon content, which leads to the d-spacing expansion. The expected expansion of the coating unit cell may be also explained by the substitution of Al into the Ti lattice positions [28]. The h-WC (ICSD 98-061-9096) and c-Co (98-062-2439) phases are the contribution of the substrate to the overall diffraction pattern of Ti-Al-C-N coatings. The expected amorphous a-C phase has not detected because of its non-crystallic character [29].

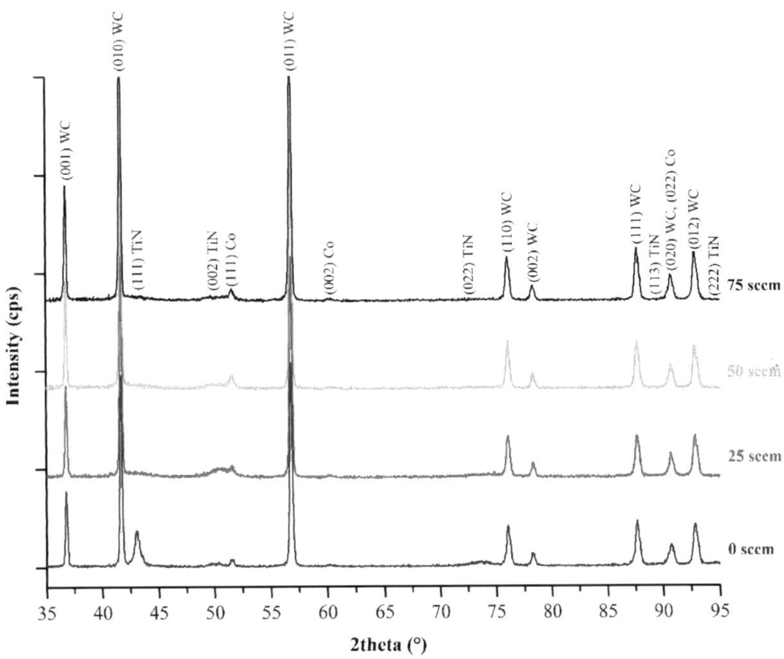

Figure 4. XRD patterns of Ti-Al-C-N coatings as a function of C_2H_2 gas flow rate.

3.4. Hardness and adhesion

Figure 5 shows the values of nanoindentation hardness (H_{IT}) of Ti-Al-C-N coatings deposited at different C_2H_2 gas flow rate. The hardness of Ti-Al-C-N coatings slightly increases as the C_2H_2 gas flow rate increased to the value of 25 sccm and reaches the maximum hardness of 33 GPa in the coating which contains 14.59 at.% C. This effect can be explained with the increase of C content in the coating, because more and more C atoms replace N atoms to form Ti(C,N) solid solution. Because of the influence of the solid solution effect, the hardness of the coatings is expected to increase. After that, the hardness values decreases with a further increase in C_2H_2 gas flow rate. Decrease in hardness at higher C_2H_2 gas flow rates is probably related with a creation of amorphous a-C phase [30]. The Young's modulus of elasticity (E_{IT}) of Ti-Al-C-N coatings is documented in the same plot in Figure 5. As C_2H_2 gas flow rate increased from 0 sccm to 50 sccm, Young's modulus of the coatings increase from 426 GPa to 441 GPa. Further increase in C_2H_2 gas flow rate from 50 sccm to 75 sccm caused a significant decrease in Young's modulus to a value 384 GPa. This decrease could be explained by the different chemical compositions in those coatings, because Young's modulus is sensitive to density and atomic structure of coating [3].

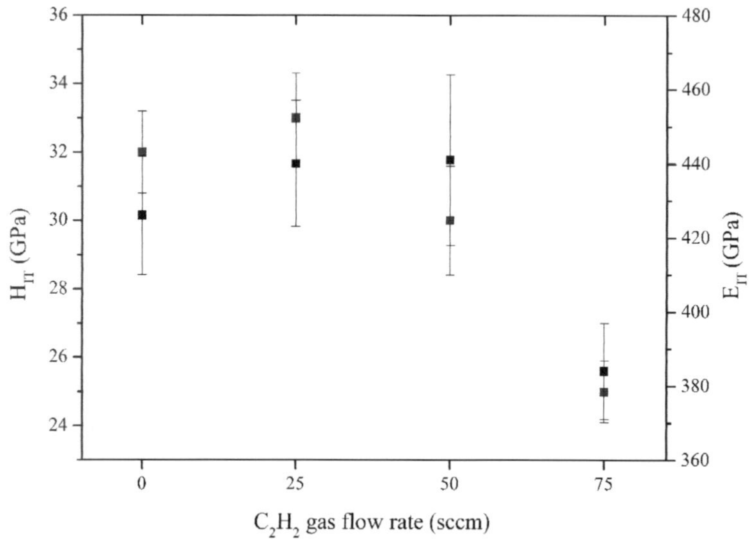

Figure 5. The effect of C_2H_2 gas flow ratio on nanoindentation hardness (H_{IT}) and Young's modulus of elasticity (E_{IT}) of Ti-Al-C-N coatings in GPa.

The adhesion of the Ti-Al-C-N coatings was evaluated with Rockwell C indentation test. Character of cracks and failures was classified according to the VDI 3198 standard. Surprisingly, the carbon free coating deposited without C_2H_2 gas was characterized by presence of cracks and cohesive failure around the indentation mark and it can be classified in the failure mode HF4 (Figure 6a). This coating still belongs to the group of acceptable adhesion level.

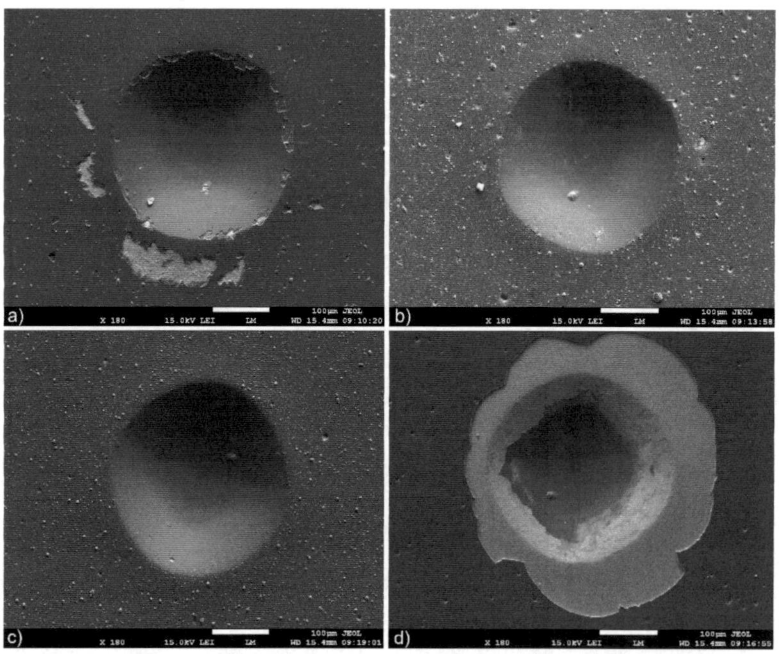

Figure 6. Failure modes of indentation marks of Ti-Al-C-N coatings deposited at different C_2H_2 gas flow rates, a) 0 sccm, b) 25 sccm, c) 50 sccm and 75 sccm.

The coatings deposited with increased C_2H_2 gas flow rates (25 and 50 sccm) were classified as HF1, because no cracks around the indentation marks were observed (Figures 6b and 6c). According to the standard, these results are considered as an excellent adhesion. However, an increase in C_2H_2 gas flow rate caused a significant deterioration of the adhesion behaviour of the coatings. In this case, a strong delamination of the coating associated with the failure mode HF6 was documented (Figure 6d).

3.5. Tribological properties

The friction coefficients curves of the Ti-Al-C-N coatings with different C_2H_2 gas flow rates are given in Figure 7. The average friction coefficient of coatings considerably decreased from 0.80 to 0.35 with increasing C content. The reduction in friction coefficient is probably related to the formation of the amorphous phase (a-C). Amorphous carbon phase exhibits low friction coefficient and from this point of view can be used to as a solid lubricant [31].

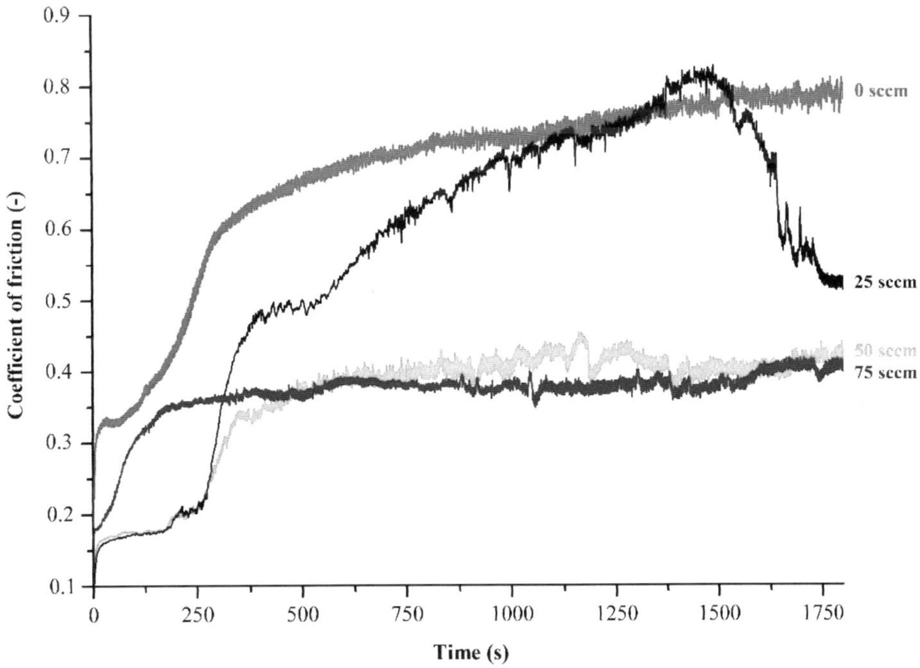

Figure 7. Coefficient of friction (COF) as a function of C_2H_2 gas flow rate of Ti-Al-C-N coatings.

4. Conclusions

The Ti-Al-C-N coatings were deposited on cemented carbide substrates by lateral rotating cathodes (LARC®) process. The effect of C_2H_2 gas flow rate on elemental and phase composition, deposition rate, cross-sectional and surface morphology, mechanical and tribological properties of the coatings was studied. As the C_2H_2 gas flow increased, the C content in the coatings increased from 0 at.% to 18.72 at.%, while N content decreased from 54.28 at.% to 47.34 at.%. All deposited Ti-Al-C-N coatings were characterized by the columnar structure of their fracture surfaces. Higher deposition rate (from 0.029 μm/min to 0.052 μm/min) was documented with increasing of C_2H_2 gas flow rate. The maximum coating thickness of ~4.71 μm was observed at C_2H_2 gas flow of 75 sccm. The presence of micro-droplets on the surfaces of all coatings was documented as well. The Ti-Al-C-N coatings consisted of a cubic B1-NaCl type $Ti_1N_{0.9}$ phase with a maximum intensity of (111) diffracting plane at a Bragg angle of 42.876° 2Θ. The peaks position shift to a lower diffraction angles is caused by increase of carbon content, which led to the d-spacing expansion. The hardness of Ti-Al-C-N coatings slightly increased as the C_2H_2 gas flow rate increased to the value of 25 sccm and reached the maximum hardness of 33 GPa, which can be explained by the increase of C content in the coating, because more and more C atoms replace N atoms to form Ti(C,N) solid solution. After that, the hardness values decreased with a further increase in C_2H_2 gas flow rate. Decrease in hardness at higher C_2H_2 gas flow rates is probably related with a creation of amorphous a-C phase. The best adhesion (HF1) exhibited the coatings deposited with increased C_2H_2 gas flow rates (25 and 50 sccm). The average friction coefficient of coatings considerably decreased from 0.80 to 0.35 with increasing C content. The reduction in friction coefficient is probably related to the formation of the amorphous phase a-C, which can be used as a solid lubricant.

Acknowledgments

This work was supported by the Project of VEGA Grant Agency of the Ministry of Education, Science, Research and Sport of the Slovak Republic and Slovak Academy of Sciences, No. 1/0540/19: "Research of possibilities of increasing the thermal and oxidation stability of Al-Ti-N based hard coatings" and also by the Slovak Research and Development Agency with grant number APVV-15-0168: "Research of modification of phase interfaces in the coating/substrate system to increase the adhesion of hard coatings". The article was also written due to the support under the Operational Program Research and Innovation for the project: "Independent research and development of new superhard coatings and their characterization by advanced experimental techniques" (ITMS code: NFP313010T598), co-financed by the European Regional Development Fund.

References

[1] Zhang K, Deng J, Guo X, Sun L and Lei S 2018 *Int. J. Refract. Met. H.* **72** 292-305
[2] Ma L W, Cairney J M, Hoffman M J and Munroe P R 2006 *Surf. Coat. Tech.* **200** 3518-3526
[3] Gu J, Li L, Miao H, Xu Y, Xu Y, Sun J, Wang X and He Z 2019 *Surf. Coat. Tech.* **365** 200-207
[4] Wang Z, Zhang D, Ke P, Liu X and Wang A 2015 *J. Mater. Sci. Technol.* **31** 37-42
[5] Kumar T S, Jebaraj A V, Shankar E, Tamiloli N, Sivakumar K 2019 *Surf. Interface.* **15** 256-264
[6] Ma S L, Ma D Y, Guo Y, Xu B, Wu G Z, Xu K W and Chu P K 2007 *Acta Mater.* **55** 6350-6355
[7] Teppernegg T, Czettl C, Michotte C and Mitterer C 2018 *Int. J. Refract. Met. H.* **72** 83-88
[8] Wang S Q, Chen L, Yang B, Chang K K, Du Y, Li J and Gang T 2010 *Int. J. Refract. Met. H.* **28** 593-596
[9] Jiang W, Wang L and Wang X 2018 *Nucl. Instrum. Meth. B.* **436** 63-67
[10] Hong D, Niu Y, Li H, Zhong X, Tu W, Zheng X and Sun J 2019 *Surf. Coat. Tech.* **374** 181-188
[11] Yoo Y H, Le D P, Kim J G, Kim S K, Vinh P V 2008 *Thin Solid Films* **516** 3544-3548
[12] Mayrhofer P H, Hörling A, Karlsson L, Sjölén J, Larsson T, Mitterer C and Hultman L 2003 *Appl. Phys. Lett.* **83** 2049-2051
[13] Li P, Chen L, Wang S Q, Yang B, Du Y, Li J and Wu M J 2013 *Int. J. Refract. Met. H.* **40** 51-57
[14] Elmkhan H, Mahboubi F, Abdollah-Zadeh A and Sabour Rouhaghdam A R 2018 *J. Manuf. Process.* **32** 873-877
[15] Gong M, Chen J, Deng X and Wu S 2017 *Int. J. Refract. Met. H.* **69** 209-214
[16] Hörling A, Hultman L, Odén M, Sjölén J and Karlsson L 2002 *J. Vac. Sci. Technol. A.* **20** 1815-1823
[17] Knutsson A, Johansson M P, Persson P O Å, Hultman L and Odén M 2008 *Appl. Phys. Lett.* **93** 14-16
[18] Chen L, Du Y, Mayrhofer P H, Wang S Q and Li J 2008 *Surf. Coat. Tech.* **202** 5158-5161
[19] Paldey S and Deevi S C 2003 *Mat. Sci. Eng. A.* **342** 58-79
[20] Kawata K, Sugimura H and Takai O 2001 *Thin Solid Films* **390** 64-69
[21] Lackner J M, Waldhauser W, Ebner R, Keckés J and Schöberl T 2004 *Surf. Coat. Tech.* **177-178** 447-452
[22] Stueber M, Barna P B, Simmonds M C, Albers U, Leiste H, Ziebert C, Holleck H, Kovács A, Hovsepian P and Gee I 2005 *Thin Solid Films* **493** 104-112

[23] Hsu Ch H, Lee Ch Y and Lee Ch Ch 2009 *Thin Solid Films* **517** 5212-5215
[24] Al-Bukhaiti M A, Al-Hatab K A, Tillmann W, Hoffmann F and Sprute T 2014 *Appl. Surf. Sci.* **318** 180-190
[25] Chim Y C, Ding X Z, Zeng X T and Zhang S 2009 *Thin Solid Films* **517** 4845-4849
[26] Sebastiani M, Piccoli M and Bemporad E 2013 *Surf. Coat. Tech.* **215** 407-412
[27] Huang Y Z, Stueber M and Hovsepian P 2006 *Appl. Surf. Sci.* **253** 2470-2473
[28] Klimovich I, Komarov F, Zaikov V, Opielak M and Kołtunowicz T N 2018 *Prz. Electrotechniczn.* **94** (6) 78-81
[29] Zhang X, Jiang J, Yuqiao Z, Lin J, Wang F and Moore J J 2008 *Surf. Coat. Tech.* **203** 594-597
[30] Patscheider J, Zehnder T, Matthey J and Diserens M 2004 *Nanostructured Hard Coatings - From Nanocomposites to Nanomultilayers* In: Voevodin A A et al (eds) Nanostructured Thin Films and Nanodispersion Strengthened Coatings, *NATO Sci. Ser. II. Math.* **155** (Dordrecht: Springer)
[31] Polcar T, Kubart T, Novák R, Kopecký L and Široký P 2005 *Surf. Coat. Tech.* **193** 192-199

IOP Conf. Series: Materials Science and Engineering **726** (2020) 012014 doi:10.1088/1757-899X/726/1/012014

Impact of phase composition on quality of ASN fertilizers

Z Štrbová[1, 2], V Jorík[2] and M Kučera[1]

[1]VUCHT a.s. Research Institute of Chemical Technology, Nobelova 34, 836 03 Bratislava, Slovak Republic

[2]Slovak University of Technology, Faculty of Chemical and Food Technology, Radlinského 2101/9, 812 37 Bratislava, Slovak Republic

zstrbova@vucht.sk

Abstract. Ammonium Sulfate Nitrate (ASN) based fertilizers are very efficient fertilizers in modern agriculture. Designed to provide targeted nutrition and high yields, they combine the extended availability of Ammoniacal (NH_4) and rapid effect of Nitrate Nitrogen (NO_3) with the multiple agricultural benefits of Sulfur. Most of these products have some tendency to form agglomerates (caking) and dust during storage in a bulk of hundreds to thousands of tons. The severity of quality deterioration can be influenced by a number of factors, such as phase composition, moisture content, particle structure, mechanical strength, hygroscopic properties, product temperature, ambient conditions, storage time and pressure. ASN fertilizer is mainly a combination of the double salts 3AN·AS ($(NH_4)_5(NO_3)_3SO_4$) and 2AN·AS ($(NH_4)_4(NO_3)_2SO_4$). It is known that presence of these mixed salts has a strong impact on the quality of fertilizer. Seeing that X-ray Powder Diffraction analysis (XRPD) is commonly used to control the quality of industrially produced fertilizers, a new X-ray method has been developed. This new method is based on the determination of the ratio of the integral intensities of the diffraction lines of the 3AN·AS and 2AN·AS components. The ratio obtained by line profile analysis provides an effective screening test of the quality of fertilizer immediately after production. In accordance with ratio value it is possible to estimate quality changes of product during long-term storage.

1. Introduction

The current intensive agricultural utilization of soils requires regular and complex fertilizer application. Ammonium Sulfate Nitrate (ASN) is a granular fertilizer, which due to its simultaneous supply of Nitrogen and Sulfur considerably improves the fertility of the soil and therefore increases the growth and yield of the plants. Because ASN contains high ammonium nitrogen content it furthermore guarantees long-term and sustainable nitrogen supply and prevents the nitrogen from being washed out of the soil. In addition, ASN supports the availability of secondary nutrients like manganese, iron, and boron in the soil [1].

The physical state in which the fertilizer is produced is of considerable importance from an agronomic point of view and with regard to satisfactory handling, transport, storage and end-application in the field. Most ASN fertilizers tend to form lumps or agglomerates (caking) and dust, which is a major problem in fertilizer handling and storage. The fertilizer caking and dustiness can be affected by several factors, e.g. chemical composition, moisture content, hygroscopic properties, product temperature and storage time [2]. The various methods of reducing or eliminating caking in fertilizers basically involve production process control, storage conditions or addition of some

IOP Conf. Series: Materials Science and Engineering **726** (2020) 012014 doi:10.1088/1757-899X/726/1/012014

anticaking agents. Furthermore, it is known that the phase composition of fertilizer immediately after production has a significant effect on its quality, especially during long-term storage [3].

X-ray Powder Diffraction (XRPD) analysis is mainly used as identification and quality control tool in the fertilizer industry. The identification of particular crystalline components of the fertilizer is carried out by the method of qualitative phase analysis. This method is based on comparing the diffractogram of each crystalline fertilizer component with the reference diffractogram from the Powder diffraction file database (PDF) [4]. Quantitative phase analysis is used to determine the content of the different phases that are present in an industrial fertilizer sample after identifying each phase [5, 6]. Overall, the quantitative analysis of ASN fertilizers can be quite difficult and problematic due to the extremely complex diffraction patterns with strongly overlapping Bragg diffractions. ASN type fertilizer can usually contain a number of components, typically four or more.

As reported in the literature, the double salts 3AN·AS ($(NH_4)_5(NO_3)_3SO_4$) and 2AN·AS ($(NH_4)_4(NO_3)_2SO_4$) (AN – ammonium nitrate, AS – ammonium sulfate) together with ammonium sulfate form the major components of ASN fertilizers [7]. Significant effects of 3AN·AS and 2AN·AS content on fertilizer quality parameters were found by diffraction analysis of high and low quality industrial samples immediately after production. Standard comparative method of qualitative analysis is not suitable for samples of ASN fertilizers by reason of complex and overlapped diffraction patterns (Figure 1). It is not possible to achieve satisfactory analysis accuracy by using it. On account of this, a new X-ray method has been developed providing an effective screening test for estimating changes in fertilizer quality during long-term storage. This new method is based on the determination of the ratio of integral intensities of selected diffraction lines of 3AN·AS and 2AN·AS components.

The aim of this work was to monitor the quality of ASN fertilizer and its changes by means of the new X-ray powder diffraction method during 4 months of bulk storage with respect to the phase composition immediately after production.

2. Experimental

2.1. Materials and physical characteristics of fertilizer materials

All analyzed samples of ASN type fertilizer in this work originate in large – scale industrial production.

In the handling and storage of fertilizers, it is important to ensure that the quality is maintained right up to the point of usage: especially no caking and minimal dust content. The most important fertilizer quality parameters – caking tendency and dustiness were measured on devices that are not commercially available, but was designed special for characterization of these fertilizer properties in our laboratory. The aim of the caking test is to hold the fertilizer sample placed in a cylindrical mold under pressure during required time interval. Then the force required to break the sinter of fertilizer is measured and defined by the unit of force Newton (N). The method used to measure fertilizer dustiness is based on the determination of the dust proportion that is separated by the airflow from the fertilizer granules. Dustiness is given as a percentage (%). An accurate method for determining the amount of water in fertilizer samples is Karl Fischer (KF) titration. It is a chemical analysis procedure based on the oxidation of sulfur dioxide by iodine in a methanolic hydroxide solution. Volumetric water content of fertilizer samples was measured with 870 KF Titrino plus titrator.

2.2. X-ray method for 3AN·AS /2AN·AS ratio determination

The powder patterns of fertilizer samples was collected within the 2θ range 10–60° with a Rigaku MiniFlex 600 X-ray diffractometer equipped with scintillation counter, using β-filtered Cu Kα radiation. As mentioned the new X-ray method is based on determination of the ratio of integral intensities of selected diffraction lines of the 3AN·AS and 2AN·AS double salts. This ratio was obtained by line profile analysis of selected diffractions in the range of Bragg angles, where is the overlap with diffractions of other fertilizer components as small as possible (18.10 – 19.15 ° 2θ). The range of angles with the smallest overlap was determined using a diffractogram obtained by combining individual simulated diffractograms of all fertilizer components (Figure 1).

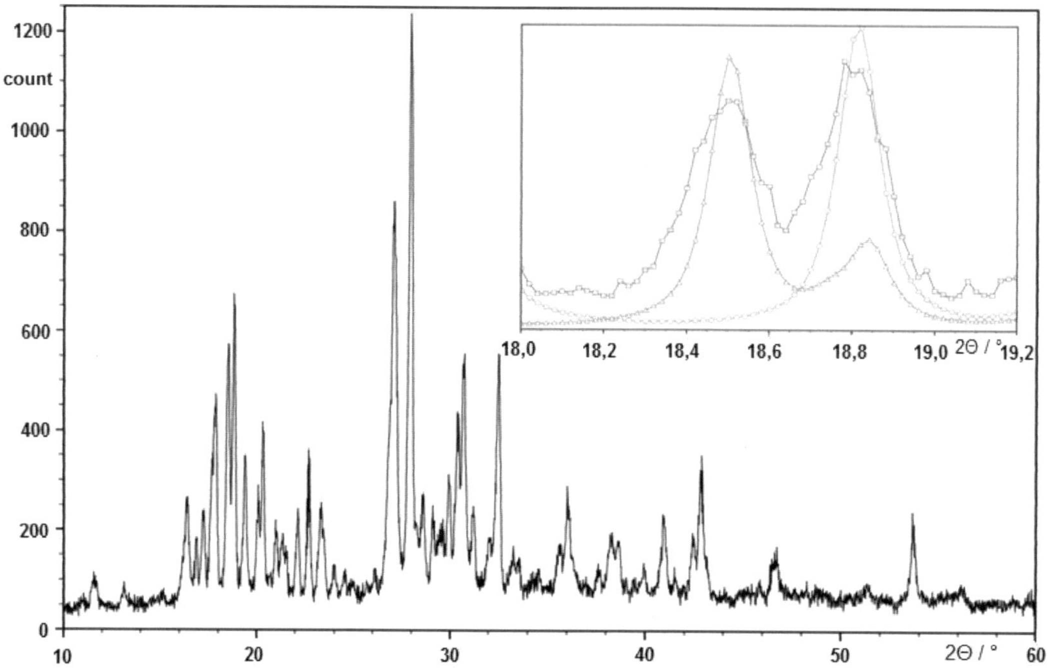

Figure 1. Diffraction pattern of industrial ASN type fertilizer. Zoomed picture in right upper corner shows the range of Bragg angles with the smallest overlap of diffractions. Black line with squares: diffractions of industrial sample, blue triangles: simulated diffractogram of 3AN·AS salt and red circles: simulated diffractogram of 2AN·AS component.

Simulated diffraction patterns of particular constituents were calculated using structure data from Inorganic Crystal Structure Database (ICSD) [8]. The X-ray method correlating the ratio of integral intensities of mixed 3AN·AS/2AN·AS salts with qualitative properties such as caking tendency and fertilizer dustiness provides an effective quality screening test immediately after production. It has been empirically found that if the ratio reaches a value less than 1.2, it can be estimated that the fertilizer will be more resistant to dramatic deterioration due to storage conditions, especially after moisture absorption. On the other hand, if the ratio value is more than 1.4, it can be expected that the metastable salt 3AN·AS will abruptly transform into stable salt 2AN·AS and qualitative parameters of fertilizer will be significantly worse [9]. The rate of such recrystallization increases with increasing relative humidity during storage. In the range of values of 1.2 - 1.4, this estimate is ambiguous, but only a minimum of samples with such a ratio value falls within the stated interval. The limit ratio values were found by gradual assignment of the ratio of integral intensities to the measured qualitative parameters of samples of real fertilizers obtained by other physical methods.

3. Results and discussion

For granular ASN fertiliser, degradation and caking is the most important problem in storage conditions. In the early research phase, a number of ASN based industrial samples were observed to deteriorate in the form of increased caking three months after production, while caking immediately after production was low (less than 20 N). The only significant difference found between them was the 3AN·AS/2AN·AS integral intensity ratio obtained by XRPD analysis. Results of XRPD analysis, caking test and water content analysis by Karl Fischer titration immediately after production and then after 3 months of storage in laboratory conditions (temperature 23 – 25 °C, relative humidity 40 – 60 %) are shown in Table 1.

Table 1. Results of XRPD analysis, caking test and water content determination

Sample	3AN·AS/2AN·AS ratio	Caking tendency (N)	Water content (%)
Sample 1 after production	1.95	15.9	0.92
Sample 1 after three months	1.55	87.1	1.06
Sample 2 after production	1.64	6.0	0.62
Sample 2 after three months	1.45	71.2	0.93
Sample 3 after production	1.10	0.0	0.82
Sample 3 after three months	0.98	60.6	0.99
Sample 4 after production	0.95	0.0	0.69
Sample 4 after three months	0.80	57.8	0.94

It can be seen from Table 1 that in the final stored product the phase conversion, depending on the moisture content and initial ratio value caused an increasing tendency to caking. Values of ratio 3AN·AS/2AN·AS decreased after three months of storage, which means that metastable salt 3AN·AS has partially transformed into the more stable salt 2AN·AS. The degree of such transformation depends on initial ratio value and as above mentioned on water content of sample [10]. Samples with a higher value of ratio immediately after production (Sample 1 and 2) have slightly more considerable tendency in caking than samples with a lower ratio after production. It can be said that samples with a lower ratio values are more stable in terms of phase transitions and consequently caking tendency.

Another series of experiments continued with the testing of industrial grade ASN fertilizers with three different 3AN·AS/2AN·AS ratios (Table 2 – 4). Thereafter these three products were stored in large piles (hundreds of tons) separately with respect to 3AN·AS/2AN·AS ratio value. During four months of storage, tests were carried out to investigate the effect of phase composition and its changes on fertilizer quality parameters, in particular the tendency of caking during long-term storage. Samples were taken monthly from each pile and the caking, dustiness, water content and 3AN·AS/2AN·AS ratio were monitored.

Samples in Tables 2 – 4 are labelled as follows: the numbers 1 – 3 indicate the average value of the 3AN·AS/2AN·AS ratio immediately after production. Letters A and B describe the origin of the sample from storage test – A means that sample was taken from the upper layers of the pile, B indicates that sample was from the lower layers of the pile. The values of each parameter shown in Table 2 are averaged out of five parallel measurements.

The moisture content of the product is usually considered to be the most important factor in promoting caking [11]. Tables 2 – 4 shows that the moisture uptake during storage is most significant in top layers of piles. It is a natural phenomenon that comes mostly from ambient moisture in warehouse. With prolonged exposure of the material to elevated humidity, caking can become quite deep. Although the water content of the samples from each fertilizer pile is roughly the same after four months of storage, the tendency to caking is very different.

Table 2. Results of XRPD analysis and tests for caking, dustiness and water content in industrial ASN fertilizer with average value 3AN·AS/2AN·AS = 1,11 immediately after production during four months of storage.

1. Average value of 3AN·AS/2AN·AS ratio = 1.11 immediately after production	1st month		2nd month		3rd month		4th month	
	1-A	1-B	1-A	1-B	1-A	1-B	1-A	1-B
3AN·AS/2AN·AS ratio	0.90	1.06	0.46	1.13	0.49	1.08	0.46	1.04
Water content (%)	1.48	1.44	1.95	1.38	1.96	1.24	2.08	1.45
Caking tendency (N)	31.1	18.3	21.5	15.0	16.9	12.1	19.6	17.5
Dustiness (%)	0.23	0.23	0.44	0.23	0.40	0.24	0.35	0.18

Table 3. Results of XRPD analysis and tests for caking, dustiness and water content in industrial ASN fertilizer with average value 3AN·AS/2AN·AS ratio = 1,27 immediately after production during four months of storage.

2. Average value of 3AN·AS/2AN·AS ratio = 1.27 immediately after production	1st month		2nd month		3rd month		4th month	
	2-A	2-B	2-A	2-B	2-A	2-B	2-A	2-B
3AN·AS/2AN·AS ratio	1.54	1.31	1.08	1.25	1.02	1.39	0.67	1.44
Water content (%)	1.12	0.91	1.86	1.29	1.94	1.21	1.95	1.32
Caking tendency (N)	24.5	23.3	13.3	25.0	6.4	14.8	73.7	19.9
Dustiness (%)	0.35	0.30	0.46	0.28	0.58	0.53	0.80	0.30

Table 4. Results of XRPD analysis and tests for caking, dustiness and water content in industrial ASN fertilizer with average value 3AN·AS/2AN·AS ratio = 2,25 immediately after production during four months of storage.

3. Average value of 3AN·AS/2AN·AS ratio = 2,25 immediately after production	1st month		2nd month		3rd month		4th month	
	3-A	3-B	3-A	3-B	3-A	3-B	3-A	3-B
3AN·AS/2AN·AS ratio	1.78	2.10	0.84	2.07	0.50	1.32	0.51	1.66
Water content (%)	1.23	1.05	1.80	1.19	1.83	1.41	2.02	1.48
Caking tendency (N)	34.7	40.0	10.0	23.9	4.7	31.0	352.1	20.5
Dustiness (%)	0.19	0.14	0.97	0.19	1.25	0.28	1.53	0.19

However, it has been found that the 3AN·AS/2AN·AS phase ratio also has a significant effect on the fertilizer caking tendency. Overall, the fertilizer with the lowest 3AN·AS/2AN·AS ratio after production (1) has the lowest caking tendency after four months storage test. On the other hand, severe caking of the top layers of pile was observed for the fertilizer with highest 3AN·AS/2AN·AS ratio after production (3-A).

Generally, increasing relative humidity during storage accelerate the 3AN·AS → 2AN·AS transformation which results in a lower 3AN·AS/2AN·AS ratio (Figure 2b). It has been found that higher relative humidity values promote the adsorption of water vapour onto the crystal surface, which enhances ion mobility [12]. The rate of 3AN·AS → 2AN·AS transformation is not only affected by the humidity but as well by the initial ratio value after production. It can be assumed that high 3AN·AS/2AN·AS ratio values immediately after production will lead to abrupt 3AN·AS → 2AN·AS transformation due to the moistening of product in storage conditions. The low value of the 3AN·AS/2AN·AS ratio immediately after production causes that recrystallization to be gradual and likely not to cause severe caking (Figure 2a).

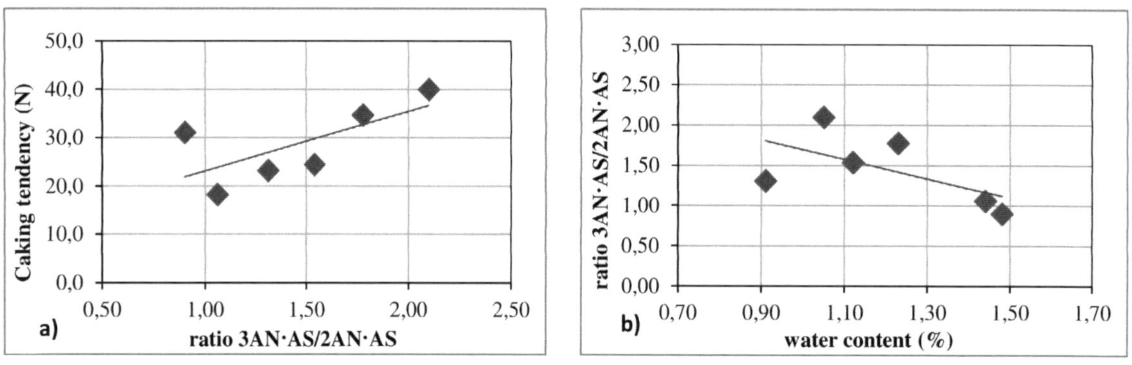

Figure 2. Relation between caking tendency and 3AN·AS/2AN·AS ratio (a) and influence of water content on ratio (b) of samples taken after 1 month of storage.

Dustiness together with caking is an important factor negatively affecting fertilizer quality. Products that generate dust make fertilizer handling and storage difficult. A larger surface of the dust particles can lead to a greater susceptibility to wetting, which in turn increases the probability of caking and reduces the shelf life. Hence, this quality parameter is given along with caking.
Figure 3 shows that caking of fertilizer is connected with 3AN·AS/2AN·AS ratio values due to very similar trends during storage. Further, it can be observed that water content influences the dustiness values rather than 3AN·AS/2AN·AS ratio. Results after 4th month of storage represent an exception to these observations, considering the strong increase of caking in upper layers of piles with higher ratio values after production (2-A and 3-A).

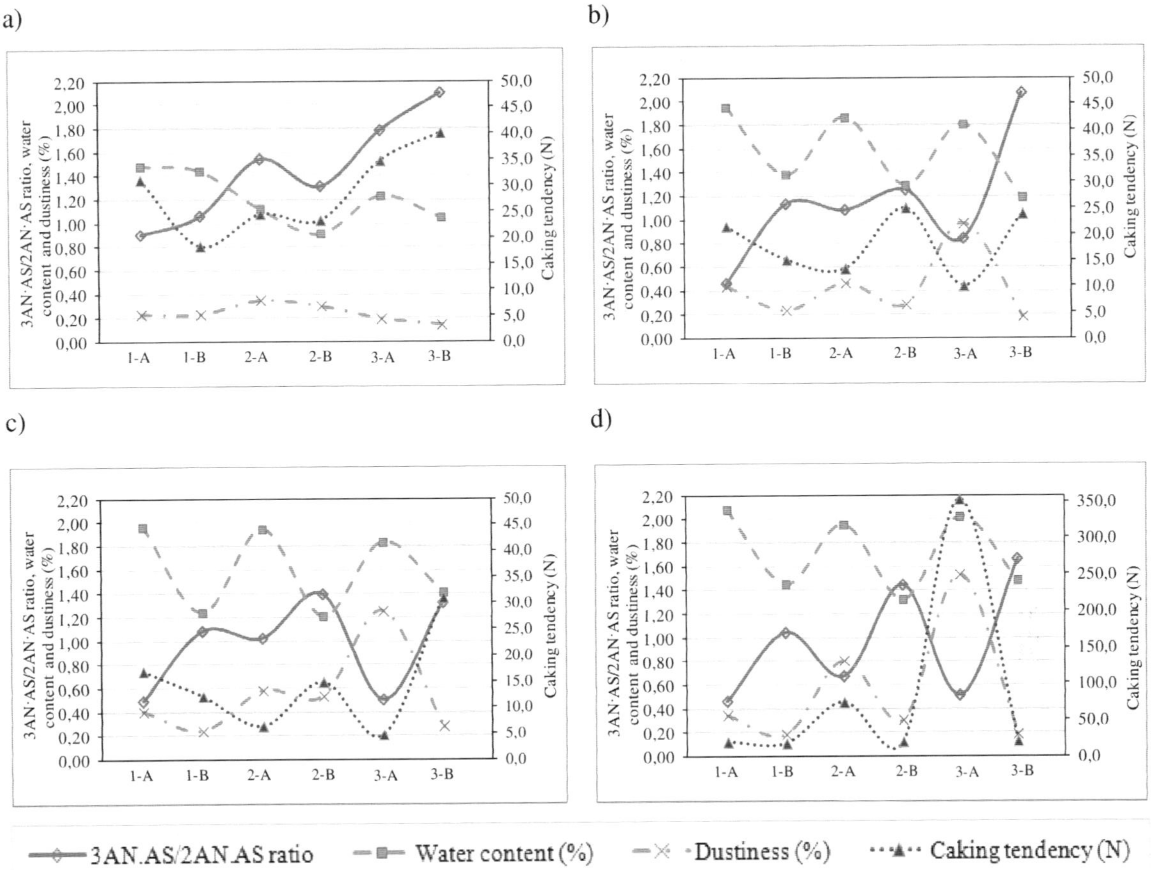

Figure 3. Trends in 3AN·AS/2AN·AS ratio, water content, dustiness an caking of fertilizer samples during storage test after 1st month (a), 2nd month (b), 3rd month (c) and 4th month (d).

4. Conclusion

This research about ASN type fertilizer indicates that caking and degradation problems of product are related to the phase composition of fertilizer immediately after production. It was found that standard search/match XRPD analysis is not suitable for this type of fertilizer because of very complex diffraction patterns. New X-ray method based on the determination of the ratio of the integral intensities of the diffraction lines of the 3AN·AS and 2AN·AS components of fertilizer was evolved with respect to all present phases. This method appears to be very fast, effective and reliable tool in screening tests of fertilizer quality immediately after production. It offers a possibility to estimate the undesirable quality changes related to phase transformation during storage. It should be noted that measured parameters can be affected by sample inhomogeneity in pile due to the large volumes of produced fertilizer, morphology of crystallites in the sample and by some other production and storage conditions.

Acknowledgement
This research was supported by Slovak Research and Development Agency grant APVV-18-0016 and VEGA 1/0639/18.

References

[1] Chien S H, Singh U, Gearhart M M and Austin E R 2013 *Soil Sci* **178** 79 – 86
[2] Rutland D W 1991 *Fertilizer Research* **30** 99 – 114
[3] Bates L 2013 *The PTF Newsletter* **18** 1– 3
[4] ICDD (2019). PDF-2 2020. International Centre for Diffraction Data, Newtown Square, PA, USA
[5] Garcia-Granda S and Montejo-Bernardo J 2013 *Reference Modul in Chemistry, Molecular Sciences and Chemical Engineering* Elsevier
[6] Pecharsky V K and Zavalij P Y 2009 *Fundamentals of Powder Diffraction and Structural Characterization* (New York: Springer Science + Business Media, Inc.) 371 – 390
[7] Babkina T S, Golovina N B, Bogachev A G, Olenev A V, Shevelkov A V and Uspenskaya I A 2012 *Russ. Chem. Bull.* **61** 33 – 39
[8] Hellenbrandt M 2004 *Crystallogr. Rev.* **10:1** 17 – 22
[9] Gezerman A O 2011 *Int. J. Chem.* **3** 123 – 133
[10] Ling T Y and Chan Ch K 2007 *Environ. Sci. Technol.* **41** 8077 – 8083
[11] Zafar U, Vivacqua V, Calvert G, Ghadiri M and Cleaver J A S 2017 *Powder Technology* **313** 389 – 401
[12] Montejo-Bernardo J, Garcia-Granda S and Fernández-González A 2011 *Acta Cryst.* **A67** 440

IOP Conf. Series: Materials Science and Engineering **726** (2020) 012015 doi:10.1088/1757-899X/726/1/012015

Stress-strain analysis of hybrid joints using numerical simulation

Š Vrtiel, M Naď and M Behúlová

Slovak University of Technology in Bratislava, Faculty of Materials Science and Technology in Trnava, Jána Bottu č, 2781/25, Trnava, 917 24, Slovakia

stefan.vrtiel@stuba.sk

Abstract. The design of joints between components of technical structures or equipment is a very important part of the design process and therefore significant attention must be paid to them. However, in certain cases, the values of the stress-strain state in a fixed joint exceed the strength limits, and it is therefore necessary to apply a combination of methods for a permanent joint, the so-called hybrid joint, which will lead to a reduction of evaluated stress-strain values. Problems of stress-strain state of hybrid permanent joints are analysed in this paper. Presented hybrid joint is formed by the combination of the pressed fit connection and fillet weld. The dimensional parameters and material properties each of type of connections are important for these hybrid joints application. Important parameters that affect the stress-strain state for this type of joint are mainly the size of the overlap of the pressed joint and also the geometric parameters of the fillet weld. In the framework of computer simulations and strength analyses are investigated the effects of different size of overlap of pressed fit joint and various design solutions of fillet weld joint on load capacity of hybrid connection under load. The results and some new knowledge resulting from these analyses are applicable in the design of similar types of joints in technical equipment and structures.

1. Introduction

The design of various technical devices and structures for which the functionality and reliability is required is subjected to the application of appropriate calculation procedures and standards [1-2]. Computer simulations and analyses based mainly on the finite element method have a leading position in predicting the stress-strain state that occurs in the proposed structure due to the applied load. The design of mutual connections between components of technical structures or equipment is a very important part of design process and therefore the significant attention must be paid to the connection design.

Generally, the components of structures can be connected by demountable or permanent joints. In order to increase the load-bearing capacity of the permanent joints, a suitable combination of joints based on different physical principles can be used. The hybrid connection presented in this paper is based on combination of press fit connection and fillet weld. The dimensional parameters as well as the material properties of each type of partial joint, used to create the hybrid joints, are important for the practical application of these hybrid joints and for determining their load-bearing capacity. Important parameters that affect the stress-strain state and load-bearing capacity of hybrid connections of these types, are mainly the overlap size of pressed fit connection and also the type and geometrical parameters of the fillet weld. Problems of stress-strain state of permanent hybrid joint are analysed in this paper. In the framework of performed computer simulations and analyses, the effects of different overlap values of pressed fit and various dimensions of fillet weld on load-bearing capacity of hybrid connection under loading are investigated.

2. Structure of permanent hybrid joint

The permanent hybrid joint of structural elements is required which consists of a cylindrical pin and wall (Figure 1). This permanent hybrid joint is created by applying two types of connections, which are based on different physical principles - pressed fit connection and fillet weld. The use of both types of joints does not allow the pin to be pulled out of the hole. It is clear that each of considered connections has certain strength limits. If only the one type of the considered connections is used (i.e., pressed fit connection or fillet weld), at a certain load level F, the destruction of the joint occurs. Thus, it can be expected that by simultaneously using both considered types of joint, the load-bearing capacity of the joint will be increased.

The geometry of structure joint and the individual dimensions of the structural elements are shown in Figure 1. The basic dimensions of the permanent hybrid joint are: pin radius - R_0, flange radius - R_p, flange thickness - L_p, pin overlap on both sides of the flange - L_c, overlap pressed fit connection - Δr_0 and size of fillet weld - a_w. The basic geometric dimensions, material properties and loads used in numerical simulations are presented in Table 1.

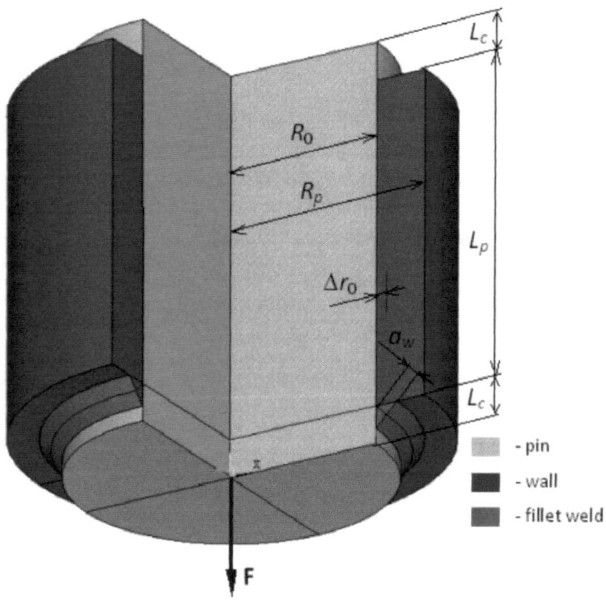

Figure 1. Structural model of permanent hybrid joint.

Table 1. The basic parameters used for numerical simulation of permanent hybrid joint.

Parameter	Symbol	Value	Unit
pin radius	R_0	30	mm
flange radius	R_p	40	mm
flange thickness	L_p	60	mm
pin overlap on both sides of the flange	L_c	7	mm
size of fillet weld	a_w	{3;4;5;6}	mm
overlap pressed fit connection	Δr_0	{1;3;5;10}	μm
axial force	F	50	kN
Young modulus	E	210	GPa
Poisson's number	μ	0.3	-
density	ρ	7800	kg.m^{-3}
friction coefficient	f	0.2	-

3. Strength evaluation of fillet weld joint under standard Eurocode 3

The design resistance of a fillet weld shall be determined using the method given in standard Eurocode 3 [1-2]. The resistance of a fillet weld may be assumed to be adequate if, at every point in its length, the resultant of all the forces per unit length transmitted by the weld does not exceed its design resistance.

The resistance of a fillet weld can be considered as satisfactory at each weld point if the resulting stresses transmitted by the weld do not exceed its design resistance expressed as follows:

- for distribution of equivalent (von Mises) stresses $\sigma_{w,eq}$ in the throat section of the fillet weld the fulfil of the following strength limit is assumed

$$\sigma_{w,eq} = \sqrt{\sigma_{\perp}^2 + 3(\tau_{\perp}^2 + \tau_{\parallel}^2)} \leq f_{uw,d}^{(1)} = \frac{f_u}{\beta_w \gamma_{Mw}} \tag{1}$$

- for distribution of normal stresses $\sigma_{\perp}$ in the throat section of the fillet weld the fulfil of the following strength limit is assumed

$$\sigma_{\perp} \leq f_{uw,d}^{(2)} = \frac{f_u}{\gamma_{Mw}} \tag{2}$$

- for distribution of maximum shear stresses τ_{max} in the throat section of the fillet weld the fulfil of the following strength limit is assumed

$$\tau_{max} \leq f_{uw,d}^{(3)} = \frac{f_u}{\beta_w \gamma_{Mw} \sqrt{3}} \tag{3}$$

where f_u - is the nominal ultimate tensile strength of the weaker part joined,
β_w - is the appropriate correlation factor,
γ_{Mw} - is the partial safety factor of welded connections ($\gamma_{Mw} = 1.25$),
$\sigma_{\perp}$ - is the normal stress perpendicular to the throat,
$\tau_{\perp}$ - is the shear stress (in the plane of the throat) perpendicular to the axis of the weld,
$\tau_{\parallel}$ - is the shear stress (in the plane of the throat) parallel to the axis of the weld.

Table 2. Parameters for assessment of bearing capacity of welded joints.

Type of steel	f_u [MPa]	β_w [-]	$f_{uw,d}^{(1)}$ [MPa]	$f_{uw,d}^{(2)}$ [MPa]	$f_{uw,d}^{(3)}$ [MPa]
S235	360	0.8	360.0	288.0	207.8
S275	390	0.8	390.0	312.0	225.2
S355	490	0.9	435.6	392.0	251.5

4. Numerical simulations and results

Computational models and required numerical simulations of permanent hybrid connections are created using the finite element method in the ANSYS program system. Since the geometry of the permanent hybrid joint as well as its loading process satisfies the conditions of axial symmetry, the computational finite element model and numerical simulations will be performed as an axially-symmetric problem. The relationship between the pin outer surface and the inner surface of hole is modelled as a contact problem. The load process is modelled using two load steps.

The first load step represents the creation of a pressed fit connection. To create a pressed fit connection, the thermoelastic analogy is applied using this and the desired overlap size pressed fit is created. The relationship between the overlap value required to create a pressed fit connection and the thermoelastic extension of pin is defined by the following equation

$$\Delta r_0 = \alpha_{pin} \cdot R_0 \cdot \Delta T \quad \Rightarrow \quad \Delta T = \frac{\Delta r_0}{\alpha_{pin} \cdot R_0} \tag{4}$$

where α_{pin} - is the coefficient of linear thermal expansion,

ΔT - is the temperature increment corresponding to ΔR_0 is applied into pin component.

The second load step is represented by the procedure of creating and considering the effect of the fillet weld and the subsequent loading of the permanent hybrid joint using the force **F** in the axial direction (see Figure 1).

Using the numerical simulations on the model defined above, stress-strain states that result from the considered load cases are investigated and analysed. The numerical simulations performed on the calculation model take into account the relevant dimensional parameters that affect the load-bearing capacity of the hybrid joint and the stress states in the throat of fillet weld.

Overview the results of stress-strain analyses for various parameters (fillet weld dimension, overlap value of pressed fit connection) are graphically presented in the Figures 2 - Figure 7.

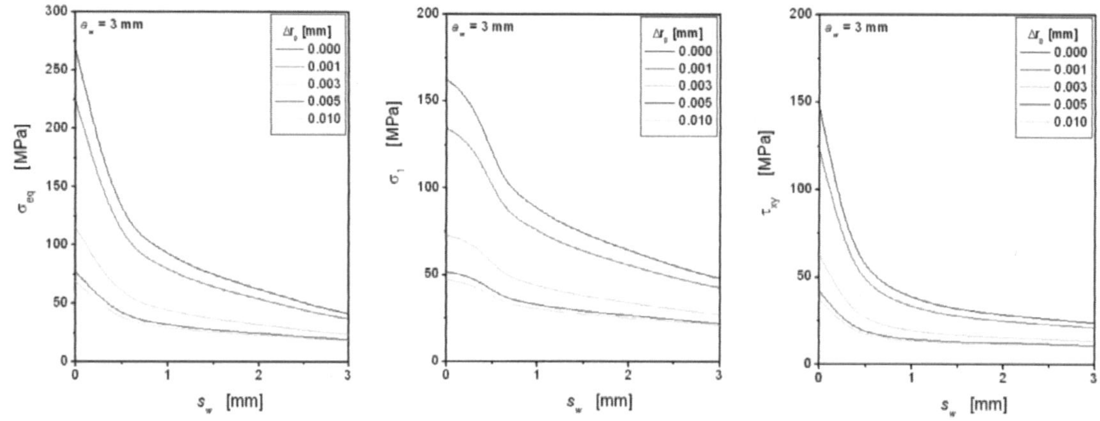

Figure 2. Dependence of evaluated stresses ($\sigma_{eq}, \sigma_1, \tau_{xy}$) in the throat of fillet weld joint (a_w = 3 mm) on overlap value Δr_0 of pressed fit connection.

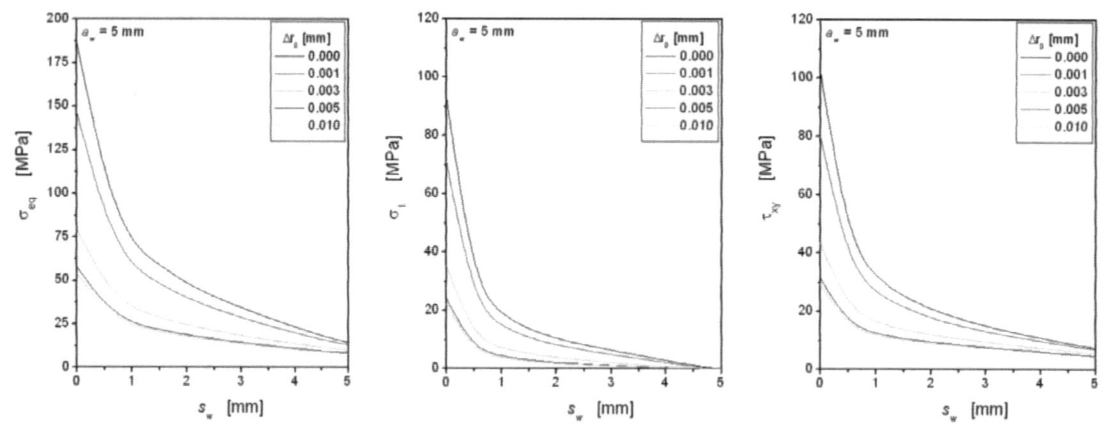

Figure 3. Dependence of evaluated stresses ($\sigma_{eq}, \sigma_1, \tau_{xy}$) in the throat of fillet weld joint (a_w = 5 mm) on overlap value Δr_0 of pressed fit connection.

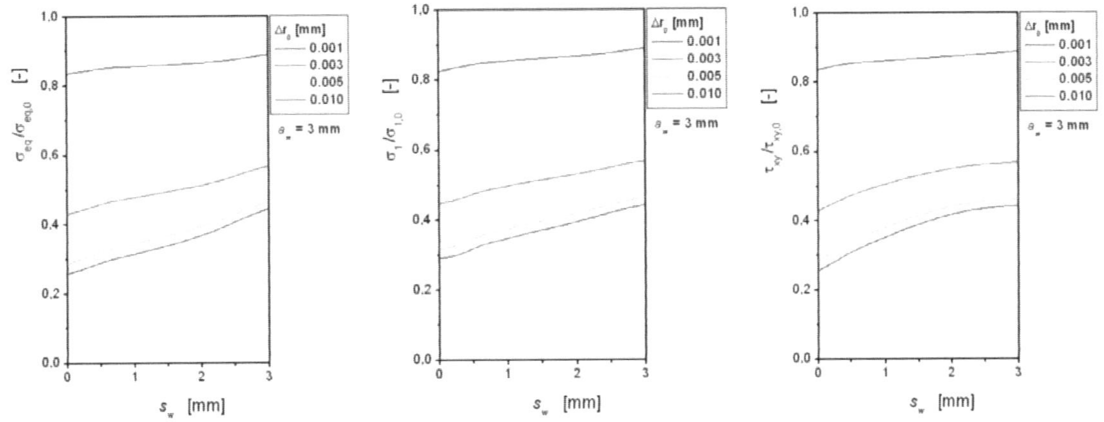

Figure 4. Influence of overlap size pressed fit connection on decrease of relative values of monitored stresses ($\sigma_{eq}, \sigma_1, \tau_{xy}$) ($a_w$ = 3 mm) in the throat of fillet weld joint.

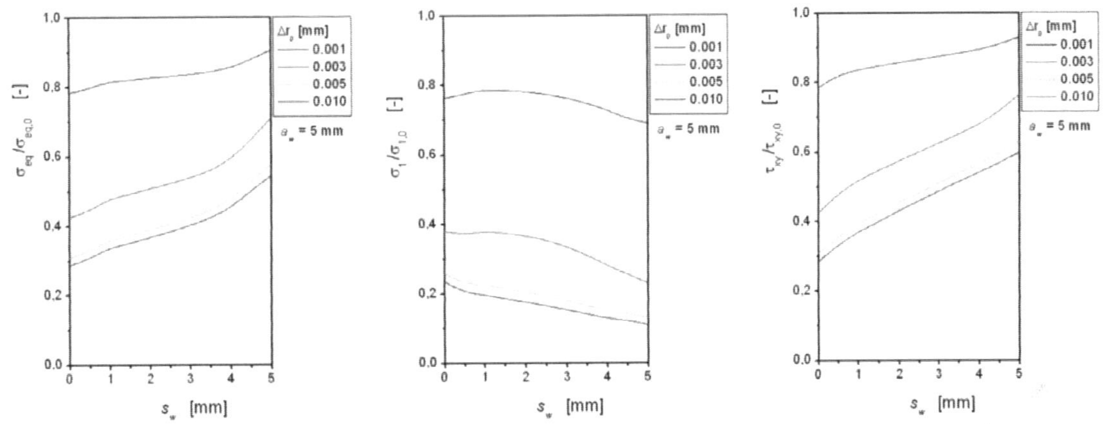

Figure 5. Influence of overlap size pressed fit connection on decrease of relative values of monitored stresses ($\sigma_{eq}, \sigma_1, \tau_{xy}$) ($a_w$ = 5 mm) in the throat of fillet weld joint.

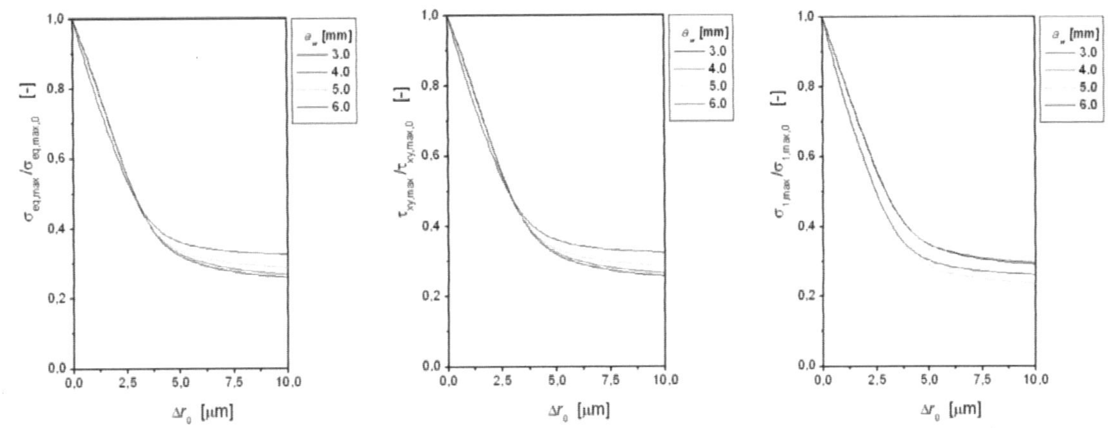

Figure 6. Influence of overlap size pressed fit connection on decrease of maximum relative values of monitored stresses ($\sigma_{eq}, \sigma_1, \tau_{xy}$) at the root of the fillet weld for various dimensions of fillet weld.

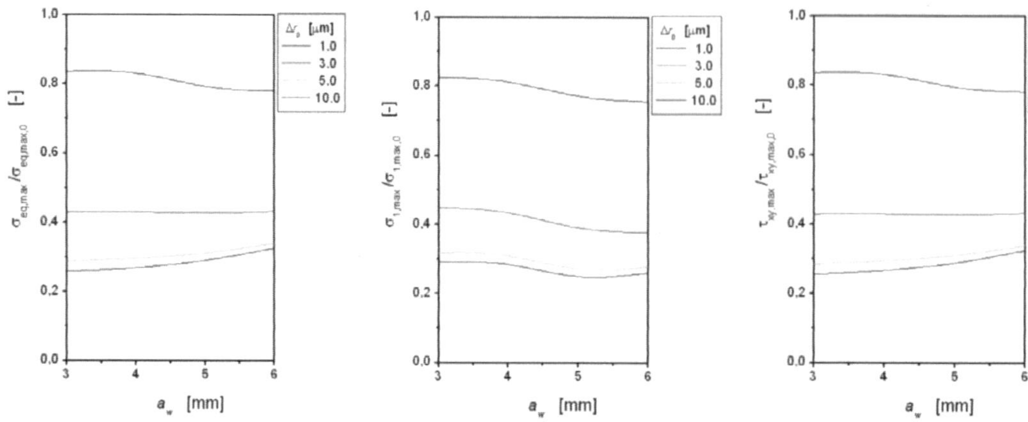

Figure 7. Influence the various dimensions fillet weld on decrease of maximum relative values of monitored stresses (σ_{eq}, σ_1, τ_{xy}) at the root of the fillet weld for different values of overlap size pressed fit connection.

Conclusions

The permanent hybrid connection, which is a combination of fillet weld joint and press-fit connection is analysed in this paper.

The main results resulting from performed numerical simulations of strength analysis permanent hybrid connection:

- enlarge overlap size for pressed fit connection causes the significant decreasing in the values of equivalent (von Mises) stress σ_{eq}, first principal stress σ_1 and also shear stress τ_{xy} in the throat plane of fillet weld,
- application of press-fit connection in permanent hybrid joint allows to use smaller dimensions of the fillet weld while the required carrying capacity of the hybrid joint is respected.

The results and new knowledge resulting from analyses of permanent hybrid joints, which confirm a significant reduction of evaluated stresses in the fillet weld, are useful and applicable in designing the similar types of joints in technical devices and structures.

References

[1] Eurocode 3: *Design of steel structures. Part 1-1: General rules and rules for buildings*. STN EN 1993-1-1, 2005.

[2] Eurocode 3: *Design of steel structures. Part 1-8: Design of joints*. STN EN 1993-1-8, 2005.

Acknowledgments

The work was supported by the Scientific Grant Agency of the Slovak Republic within the project VEGA No. 1/1010/16.

Fatigue life of thermoset composite materials

M Zaludek, S Rusnakova, M Kubisova, O Bilek and K Karvanis

Tomas Bata University in Zlín, Faculty of Technology, Vavrečkova 275, 760 01 Zlín, Czech Republic

zaludek@utb.cz

Abstract This article deals with the lifetime of laminated materials produced by different production technologies (hand lamination technology, vacuum bagging technology and prepreg technology with curing in the oven) during cyclic repeated bending stress. Like tested materials composite systems were chosen composite systems with epoxy matrix and carbon reinforcement (Kordcarbon CC200T, epoxy resin with trade name L285) and second composite epoxy system with glass reinforcement (Quadra-axial glass fabric, named Saertex Q-E-820, epoxy resin Biresin CR 82 and third prepreg systems (unidirectional prepreg, trade name Deltapreg VV430U-DT860W-39% and prepreg system with glass fabric VV320P_DT806R-37%). Fatigue tests were performed by cyclic bending loads during a three-point arrangement on a universal servo-hydraulic testing machine INSTRON 8871. Experimentally the number of cycles fracture of the material was determined at 90 % and 80 % of the maximum breaking load.

1. Introduction

Increasingly, there is a need to know the behaviour of composite materials in fatigue cyclic loading in the design and development of new products. Carbon laminates are characterised by excellent fatigue resistance. The best combination is the epoxy matrix/carbon fiber. Carbon composites with epoxy resin are very suitable material for cyclic and variable stresses because they resist well. The tensile fatigue limit of these composites is usually about 70 % of the static tensile strength; the AC tensile strength is about 35 % of the static tensile strength. Because these composites have a very favourable strength-to-weight ratio, their use is very advantageous for aircraft structures. The advantage is high fatigue strength due to static strength and its slight decrease with the number of cycles to fracture, high residual strength, low notch sensitivity, low sensitivity to loading frequency, as well as a small decrease in stiffness with the number of loading oscillations performed.

Another significant advantage is the possibility of obtaining the required stiffness, strength, Poisson number or preload by the appropriate layering of the desired layers. The disadvantages include weak compressive load resistance, squeezing and corrosion sensitivity when in contact with Al alloys and steels. A significant disadvantage is the absorption of moisture and the consequent deterioration of mechanical properties. However, by making a suitable design, these disadvantages can be overcome. Compared to metallic materials, the Wöhler carbon laminate curves are very flat, and there is no fatigue limit [1]. In connection with S-N curves, high-cycle and low-cycle fatigue should be distinguished. The first term identifies a situation of long fatigue life; that is, the voltage amplitude is sufficiently low, and thus, plastic deformation is not dominant for material behaviour.

High-cycle fatigue starts from 10^2-10^4 cycles. In addition to the position of the Wöhler curves, which determine the number of cycles failing, the residual lifetime, based on the cumulation theory of damage, is determined for a specific voltage and several cycles. The lifetime is exhausted if the sum of the individual damages is greater than 1. The advantages of using a carbon composite are mainly in low weight, high strength, excellent shock absorption properties, high durability and very favourable fatigue characteristics. A comparison of the fatigue behaviour of carbon composites with other materials is shown in figure 1 [2-3].

The aim of the paper is to determine the fatigue life of fiber composites produced by the method of manual lamination.

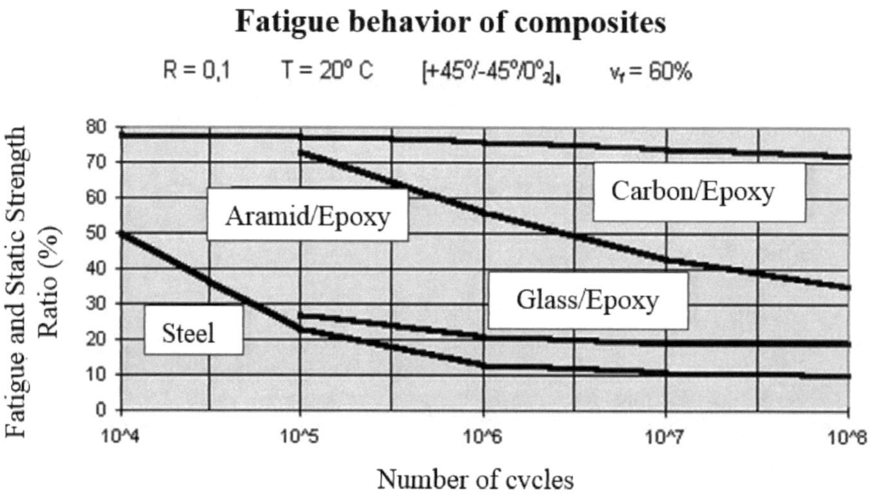

Figure 1. Fatigue behaviour of composites - comparison [3].

2. Testing methodology

Cyclic fatigue tests were performed on servo-hydraulic universal test machine INSTRON 8871, shown in figure 2. This machine is designed primarily for cyclic fatigue testing but it can be used also for static tests. The machine base is equipped with a T-groove, where a wide range of test components can be clamped for different types of tests. Fatigue tests were performed by a cyclic bending load in a three-point configuration. The distance between the supports was set the same as for the static test, 80 mm. The static bending test was carried out according to the CSN EN ISO 14 125 standards [4].

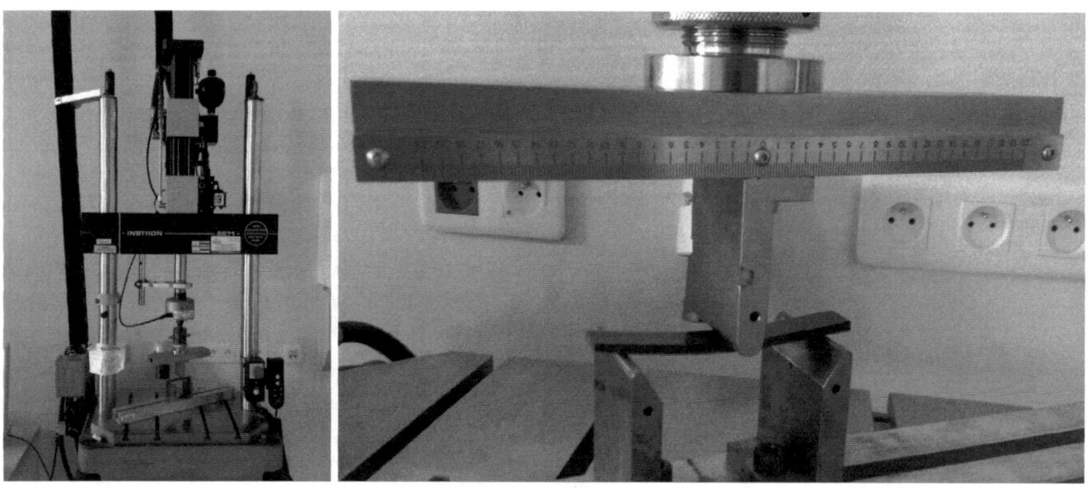

Figure 2. Cyclic bending test arrangement.

Several test measurements were made before performing the test itself. Subsequently, a load frequency of 6 Hz and two load sizes were selected - 80 % and 90 % of the maximum bending force found in the static test. Furthermore, the mean oscillation force Fs and the load amplitude Fa were selected. The load curve was sinusoidal, see figure 3.

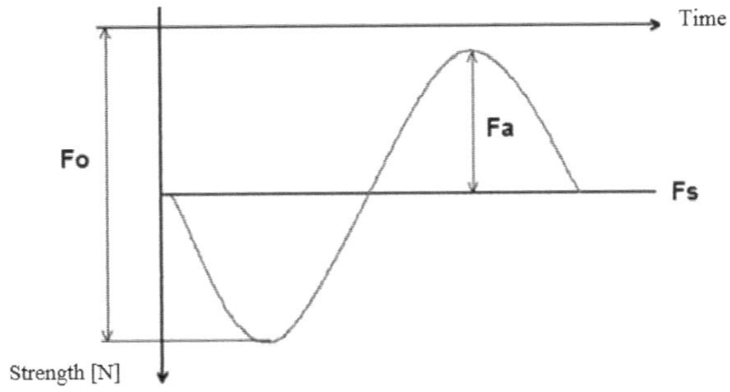

Figure 3. Cyclic loading.

3. Materials tested

Three types of materials were selected for testing, which differed in the reinforcement component and the production technology.

For fabric no. 1, carbon fabric was used as a reinforcing component from Kordárna Plus a.s., which is a direct manufacturer of these fabrics. The fibers are supplied by TORAY and bear the designation FT300B - 3000 - 40B. This fiber has a tensile strength of 3805 MPa and a tensile modulus of 232 GPa [5]. The carbon fabric used for the production of test specimens has the designation CC200 T - 100 KORDCARBON, weight 200g / m^2, twill weave, see figure 4.

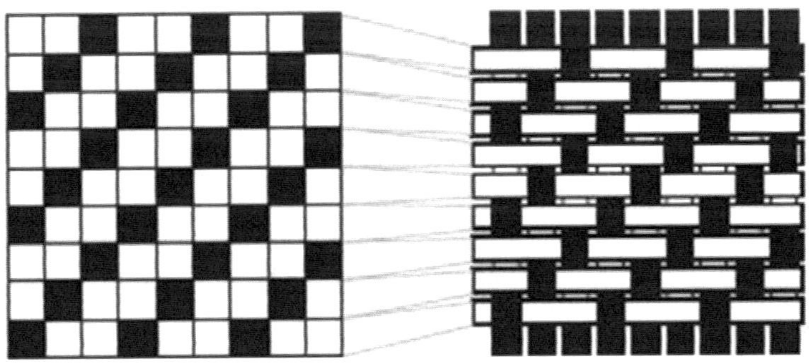

Figure 4. Twill weave pattern.

HAVEL L285 epoxy resin was used as a matrix together with H285 hardener in a 100:40 mixing ratio. This low viscosity epoxy system is designed to produce high static and dynamic load parts. It is suitable for hand lamination in combination with glass, carbon or aramid fibers.

A thin layer of the epoxy mixture was applied to the cleaned and separated form, and then the first layer of fabric was placed thereon. Subsequently, the fabric was soaked with an epoxy mixture using a brush (figure 5). A plastic spatula was used to displace any bubbles.

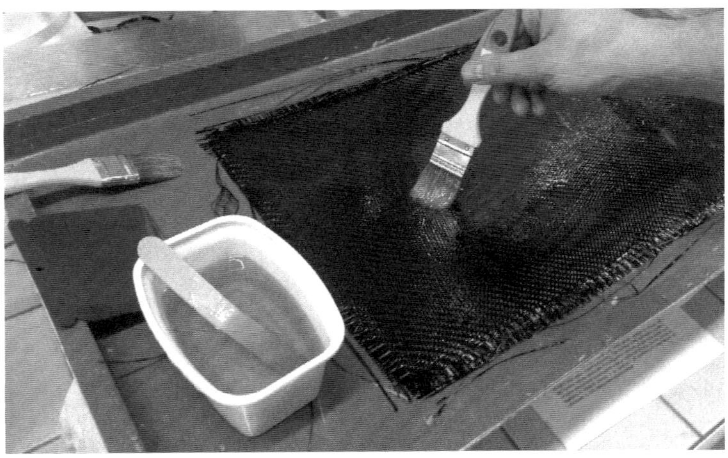

Figure 5. Production of the board by the method of manual lamination.

For the production of material no. 2, a 4-axial Saertex glass fabric was used. Q-E-820 samples consist of eight layers. The matrix was a Biresin CR 82 resin with a CH 80-1 hardener at a ratio of 100:27. These specimens were made by hand lamination as well as Kordcarbon specimens, but also vacuum compression was used, see figure 6. Test specimens for static and fatigue bending tests were machined from prepared cured plates to 20 mm ×100 mm.

Figure 6. Vacuum pressing.

For the production of material no. 3, glass prepregs (Deltapreg canvas VV320P and unidirectional UD Deltapreg VV430U) of the Italian manufacturer Delta were used. The sample plate produced consists of two types of reinforcing glass prepregs - one-way prepreg with a density of 430g / m^2 and a prepreg screen with a density of 320 g/m^2. The UD prepreg is oriented in the axis of the test specimens; the canvas prepreg is oriented at an angle of 45°. The arrangement of the individual layers is shown in figure 7. The manufacturer used Deltatech DT806 epoxy resin as the prepreg matrix.

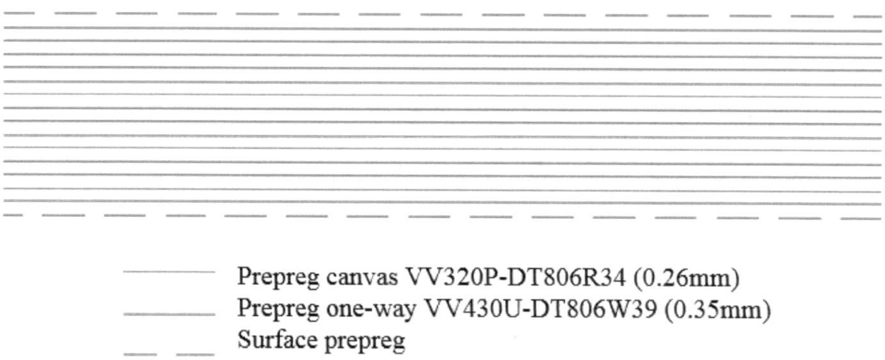

Figure 7. Arrangement of individual layers of fibreglass board

4. Results

The main result is the number of refractive cycles that the tested composites withstand the 80 % and 90 % of the static flexural strength, respectively.

4.1. Carbon Laminate Kordcarbon CC200T

At a load corresponding to 80 % of the static strength of the body, it was 150,000 cycles without failure and was subsequently subjected to a static test for residual strength.

Table 1. Comparison of average values of cyclically unloaded bodies and bodies after 150,000 cycles.

Kordcarbon CC200T	***E* [MPa]**	**σ_{fM} [MPa]**
A cyclically unloaded test specimen	38600	597
150000 Cycles 80 % σ_{fM}	32450	572

After 150,000 cycles at 80 % strength, the test specimens showed a slight decrease in elastic modulus, or more specifically, 15 %. There was also a slight decrease in the strength limit, but this decrease is at the limit of measurement error.

Table 2. Number of cycles for Kordcarbon test specimens.

test specimen number	Number of cycles	
	load 80 % σ_{fM}	load 90 % σ_{fM}
1.1	150 000*	46 250
1.2	150 000*	78 541
1.3	150 000*	23 122

In table 2, significant differences in the number of cycles that the test specimens could withstand at 90 % strength were observed. These differences can be caused by defects in the structure of the material, such as bubbles or insufficient resin saturation at some points in the sample. The quality not only of carbon but also of other composites is very dependent on the careful production.

4.2. Fiberglass 4axial fabric Saertex Q-E-820

The number of cycles after which the specimens were broken are shown in table 3.

Table 3. Number of cycles for test specimen from 4-axial glass fabric.

test specimen number	Number of cycles	
	load 80 % σ_{fM}	load 90 % σ_{fM}
1.1	12 326	3 749
1.2	11 343	2 534
1.3	15 032	-

As can be seen from table 3, 4-axial fibreglass test specimens did not withstand many cycles for such high load levels.

4.3. Glass prepreg Deltapreg VV320P + VV430U

The number of cycles after which the specimens were broken are shown in table 4.

Table 4. Number of cycles for test specimen from glass prepreg.

test specimen number	Number of cycles	
	load 80 % σ_{fM}	load 90 % σ_{fM}
1.1	85 821	12 073
1.2	108 249	20 487
1.3	102 460	19 452

Glass prepreg specimens withstood relatively high cycles compared to 4 axial glass fabric specimens. Prepreg products have a high proportion of fiber reinforcement, thus achieving excellent mechanical properties. Compared to the standard lamination technology, the fiber reinforcement is evenly saturated here, which improves fatigue resistance many times over.

5. Discussion

Deltapreg glass prepreg specimens withstood much more cycles than Saertex 4-axial fiberglass specimens. A comparison of the number of cycles is shown in figures 8 and 9.

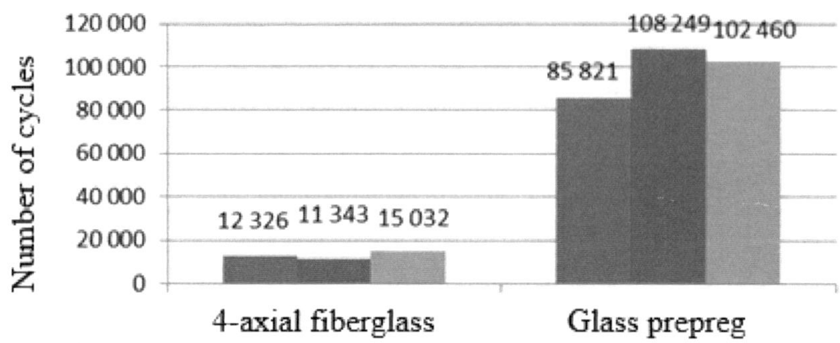

Figure 8. Comparison of the number of cycles of fiberglass under a load of 80 % σ_{fM}.

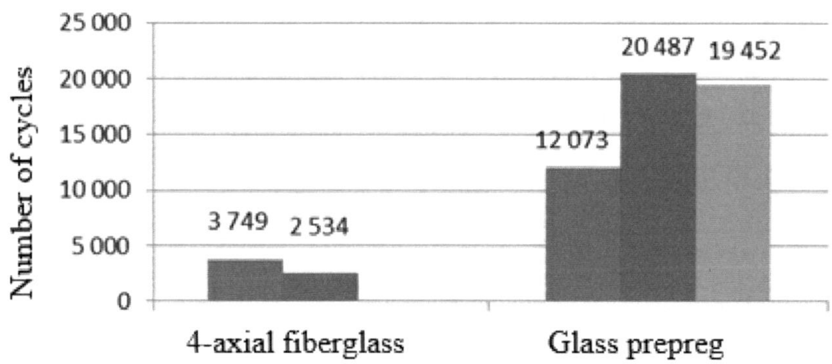

Figure 9. Comparison of the number of cycles of fiberglass under a load of 90% σ_{fM}.

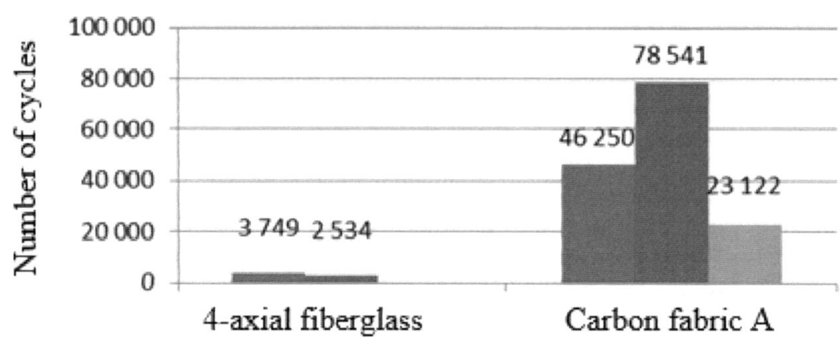

Figure 10. Comparison of the number of cycles of glass and carbon laminates under load 90 % σ_{fM}.

Glass prepreg specimens have shown a relatively good fatigue life not only against 4-axial fiberglass specimens but also compared to carbon fabric specimens. At a load of 80% of the flexural strength, these test specimens withstood on average 98.800 cycles, while the Kordcarbon CC200T carbon fabric housings withstood this load of 150.000 cycles without breaking the specimen, see figure 11. The comparison of the number of cycles at 90 % stress limit for Kordcarbon CC200T and Saaxtex 4axial fiberglass is shown in figure 10.

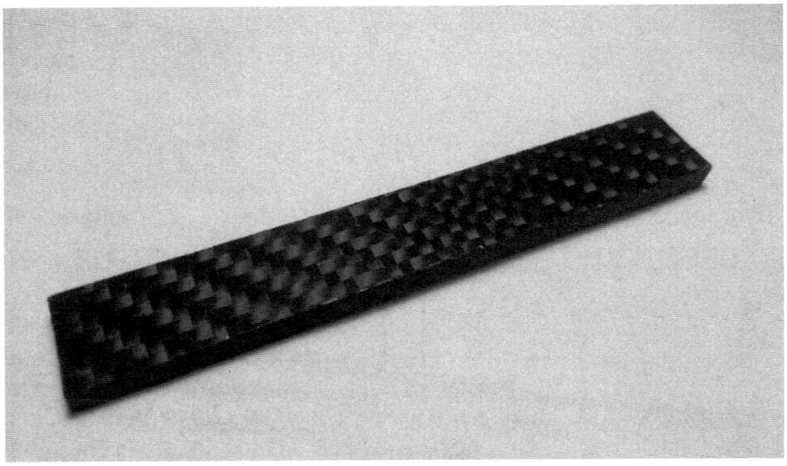

Figure 11. Kordcarbon CC200T carbon fabric test specimen after 150,000 cycles without damage.

6. Conclusion

The results of the experiments show that the carbon composite Kordcarbon CC200T achieves the highest durability. The second place in terms of durability under cyclic loading is the material made of a combination of glass prepreg Deltapreg. On the contrary, the worst material of durability under cyclic loading was the composite made of four axial glass fabric Saertex Q-E-820, produced by the method of manual lamination followed by vacuum pressing.

Acknowledgements

This work and the project was realized with the financial support of the internal grant of TBU in Zlín No. IGA/FT/2019/001 funded from the resources of specific university research.

This work and the project developed with the financial support of the internal grant of TBU in Zlin No.RVO/FT/2019/003.

References

[1] Bezazi A R, El Mahi A, Berthelot J M and Bezzazi B. (2003) Flexural fatigue behavior of cross-ply laminates: An experimental approach *Strength of materials* **35** (2) 149-161

[2] Rusnakova S, Fojtl L, Zaludek M and Rusnak V 2014 Design of material composition and technology verification for composite front end cabs *Manuf. Technol.* **4** 607 – 11

[3] Duratec.cz [online]. [cit. 2016-04-28]. Retrieved from: http://www.duratec.cz/cs/techinfo/pouzivane-materialy/uhlikova-vlakna-karbonove-kompozity/

[4] ISO 14125 : *Fibre-Reinforced Plastic Composites – Determination of Flexural Properties*

[5] T300 *Data Sheet – Toray Composite Materials America, Inc.* [online]. [2019-08-10]. Retrieved from: https://www.toraycma.com/file_viewer.php?id=4462

[6] Rusnakova S, Capka A, Fojtl L and Zaludek M 2016 Technology and Mold Design for Production of Hollow Carbon Composite Parts *Manuf. Technol.* **16** 799 - 804

[7] Javorik, J and Bilek, O 2014 Numerical analysis of bushing of car stabilizer *Int. J. Mech.* **8** 289 -97

[8] Javorik, J 2016 Numerical Optimization of Large Shade Sail Support. *Manuf. Technol.* **4** 707-12

IOP Conf. Series: Materials Science and Engineering **726** (2020) 012017 doi:10.1088/1757-899X/726/1/012017

Susceptibility to the intergranular attack in austenitic stainless steels

V Zatkalíková, L Markovičová, M Uhríčik and P Hanusová

University of Žilina, Faculty of Mechanical Engineering, Department of Material Engineering, Univerzitná 8215/1, 01026 Žilina, Slovak Republic

viera.zatkalikova@fstroj.uniza.sk

Abstract. Intergranular corrosion is very dangerous local corrosion form that often leads to the dislodgment of individual grains and to the intensive negative influence on the mechanical properties of the material. The susceptibility of the austenitic stainless steels to the intergranular corrosion is connected with their exposition in the temperature range of 500-800 °C ("critical temperatures") and with consequent slow cooling in the air which leads to the precipitation of chromium-rich carbides at the grain boundaries. This process consumes chromium from a narrow band along the grain boundaries and the steel becomes to be sensitized and susceptible to the intergranular corrosion in aggressive environments. This article deals with the susceptibility to the intergranular corrosion of three austenitic stainless steels (AISI 304, 316L, 316Ti). Both "as received" and improperly heat treated specimens (sensitization for 10 hours at 650 °C, cooling in the air) were tested by ASTM A262 standard method, A and E practices. Optical microscopy and SEM analysis were used for the assessment of the obtained results.

1. Introduction

Intergranular corrosion is the special form of local corrosion, which can occur in many alloy systems when the corrosion rate of the grain-boundary areas is higher than that of the grains interiors. The result is the corrosion of a narrow path preferentially along the grain boundaries. This localized attack is very dangerous because the surface damage of the material is usually easily overlooked but it often leads to the dislodgment of individual grains and to the intensive negative influence on the mechanical properties. Although the detailed mechanism of intergranular corrosion varies with each metal system, its physical appearance at the microscopic level is quite similar for most systems [1-4].

Intergranular corrosion is very typical corrosion attack in austenitic stainless steels. In common oxidizing environments these steels have good corrosion resistance due to the formation of passive protective surface film. However, under the action of aggressive halide ions, local breakdown of passivity occurs, causing pitting corrosion [2, 5-6]. The susceptibility of the austenitic stainless steels to the intergranular corrosion is connected with their exposition in the temperature range of 500 – 800 °C ("critical temperatures") and with consequent slow cooling in the air which leads to the precipitation of $M_{23}C_6$ chromium-rich carbides on the grain boundaries [7-9]. The precipitation of chromium carbides consumed the main alloying element - chromium from a narrow band along the grain boundary and this makes the zone anodic to the unaffected grains. If the chromium content near the grain boundaries drops under the passivity limit (11.5 wt. %), the steel becomes to be sensitized and susceptible to the intergranular corrosion in aggressive environments. The chromium depleted zone becomes the preferential path for corrosion attack or crack propagation if under tensile stress [1-4, 7]. The sensitisation temperature range is often encountered during isothermal heat treatment, slow cooling from the solution annealing temperature, improper heat treatment in the heat affected zone of the welds or weld joints or hot working of the material. Degree of the sensitisation is influenced by the factors such as the steel chemical composition, grain size, degree of strain, or temperature and time of isothermal annealing [7].

This article deals with the susceptibility to the intergranular corrosion of three austenitic stainless steels (AISI 304, AISI 316L and AISI 316Ti). Both "as received" and improperly heat treated specimens (sensitization for 10 hours at 650 °C, cooling in the air) were tested by ASTM A262 standard method, A and E practices. Optical microscopy and SEM analysis were used for evaluation of the obtained results.

2. Experimental material

AISI 304, AISI 316L and AISI 316Ti austenitic stainless steels with chemical compositions in table 1 were used as the experimental material. They were purchased in sheets (1000 x 2000 mm) of 1.5 mm thickness. Their production processes were based on continuous casting in electric arc furnace. Then they were annealed (AISI 304 at 1040 – 1100 °C, AISI 316L and 316Ti at 1050 °C). The IIB surface finish (smooth and matte metallic glossy surface) was realized by pickling after slightly smoothing rolling [10]. Microstructures of "as received" steels (figure 1) are polyhedral, created by austenitic grains with numerous twins, which could be created by annealing or by rolling.

Table 1. Chemical compositions of experimental stainless steels (wt. %).

	Cr	Ni	Mo	Ti	Mn	N	C	Si	P	S	Fe
304	18	8.01	-	-	1.40	0.075	0.027	0.38	0.031	0.0037	balance
316 L	16.79	10.14	2.03	-	0.82	0.05	0.02	0.31	0.03	0.001	balance
316 Ti	16.54	10.56	2.02	0.303	1.07	0.009	0.023	0.48	0.032	0.006	balance

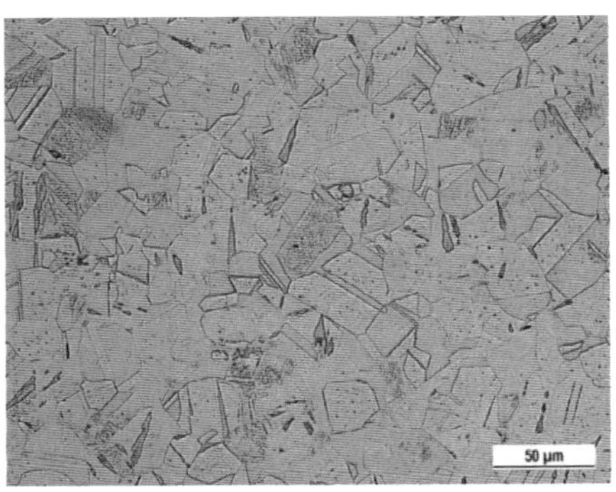

A)

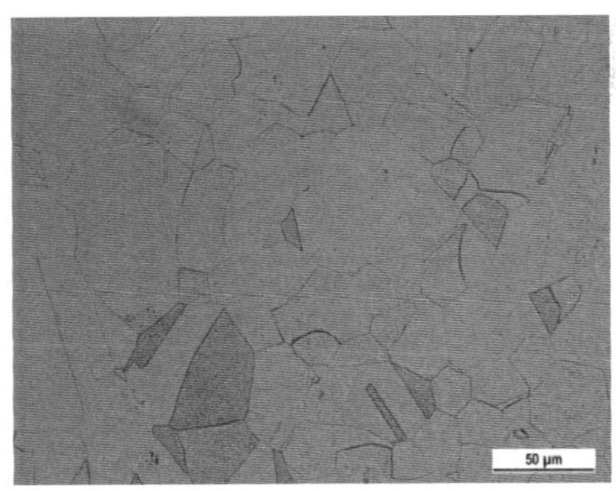

B)

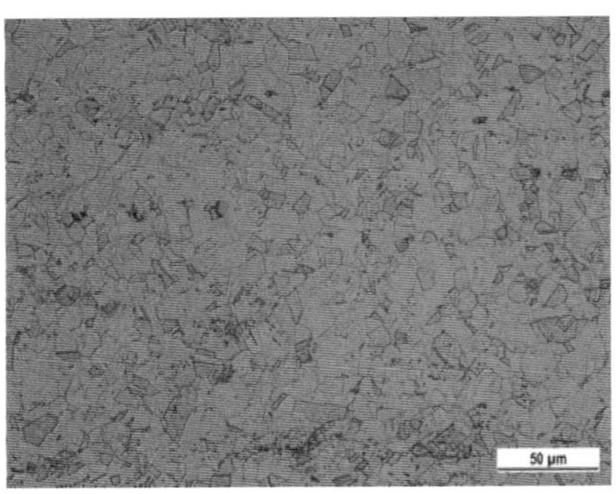

C)

Figure 1. Microstructures of the experimental stainless steels in "as received" state, longitudinal sections: A) AISI 304 (glycerine + HNO_3 + HCl etch.); B) AISI 316L (Kallings 2 etch.); C) AISI 316Ti (Kallings 2 etch.).

The steels are of different size of grains. The orange colored cuboid-shaped titanium carbides are markedly seen in AISI 316Ti microstructure (figure 1C).

3. Experimental procedure

The rectangular specimens (14 mm x 40 mm x 1.5 mm) were prepared for experiments. A part of specimens were improperly heat-treated to evoke a sensitization. Then both "as received" and heat-treated specimens were tested by ASTM A262 standard method A and E practices for an assessment of their susceptibility to the intergranular corrosion.

Heat treatment of specimens was performed in furnace at 650 °C for 10 hours with consequent slow cooling to create suitable diffusion conditions for precipitation of chromium carbides. The temperature of 650 °C was chosen according to the graph of carbon solubility in austenite [5].

The oxalic acid etch test was performed according to A practice of ASTM A262 standard method under the conditions listed in table 2 [11]. The specimen surfaces were mechanically treated by grinding (500 and 1000 mesh) before the electrochemical etching process. For the etching, each specimen was immersed into electrolyte and connected to the positive pole of the power source as an anode. A cathode was of the same material as the anode [11]. The etched surfaces (the specimen sides facing the cathode) were evaluated by the ZEISS Observer 21M microscope.

Table 2. Conditions of the electrochemical etching (ASTM A262, A practice).

Solution component	Content (wt. %)	Temperature (°C)	Current density (A.cm^{-2})	Time (s)
oxalic acid	10	20 ± 3 °C	1.0	90
demineralized water	90			

The specimens for E practice of ASTM A262 standard method (Strauss test) were metallographically prepared. Then they were placed on electrolytic copper pieces and immersed into solution. The test was carried out according the conditions listed in table 3 [11]. The specimens after Strauss test were studied by the ZEISS Observer 21M microscope and by SEM analysis (Tescan Vega microscope).

Table 3. Conditions of Strauss test (ASTM A262, E practice).

Solution component	Content (wt. %)	Temperature (°C)	Time (hours)
copper sulfate (anhydrous)	6	100 °C	15
sulfuric acid	16		
demineralized water	78		

4. Experiment results and discussion

The microstructure of specimens after heat treatment (sensitization) visualized by classical chemical etching (Kallings 2) shows no radical changes (figure 2) only some grain boundaries are highlighted compared to the "as received" state. On the AISI 316Ti specimen (figure 2C), titanium carbides are concentrated into clusters (similar clusters are not visible in the "as received" state). Due to the minimal differences between the "as received" and post-heat treatment states can be deduced that the chemical etching used did not enable to visualize a possible precipitation of chromium carbides at the grain boundaries caused by the given improper heat treatment.

The oxalic acid etch test is used in connection with other tests as a rapid method for identifying those specimens which are certain to be free of susceptibility to the rapid intergranular attack in the various hot acid tests (4 to 240 hours of exposure) [11]. The assessment of the susceptibility to the intergranular corrosion is based on the classification of grain boundaries attack [3, 7, 11] that distinguishes three types of etch structures (step, dual and ditch structures). If there are no ditches at grain boundaries only steps between grains the structure is marked as the step one. This structure type indicates the highest resistance to the intergranular corrosion.

IOP Conf. Series: Materials Science and Engineering **726** (2020) 012017 doi:10.1088/1757-899X/726/1/012017

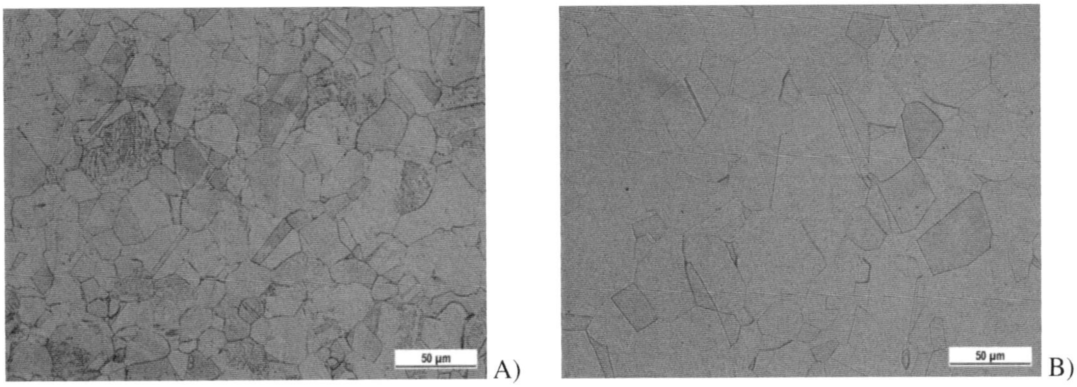

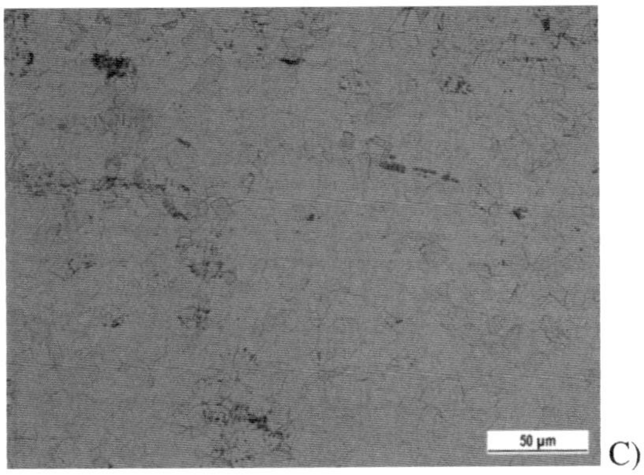

Figure 2. Microstructures of the experimental stainless steels after performed heat treatment (sensitization), areas of the specimens, (Kallings 2 etch.): A) AISI 304; B) AISI 316L; C) AISI 316Ti.

The presence of ditches indicates the dissolution of carbides at the grain boundaries during the previous etching process. When no single grain is completely surrounded by ditches, the structure is marked as the dual one. The ditch structure is reflected by one or more grains completely surrounded by ditches [3, 7, 11].

According to microstructures after the etch test (figures 3 and 4) AISI 304 and AISI 316L steels showed strong differences in intergranular corrosion susceptibility between the "as received" and the sensitized state.

The step structures in the "as received" state could be caused by the different dissolution rates of the variously oriented grains and they are probably not related to the chromium carbides precipitation. Therefore can be concluded that these steels were not exposed to critical temperatures in the production process, and thus they were not sensitized.

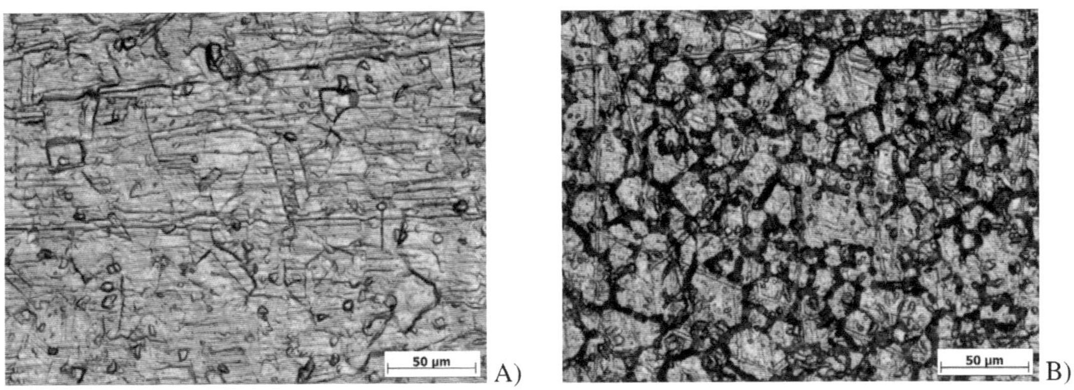

Figure 3. Microstructure of AISI 304 after the etch test: A) "as received" state; B) the state after heat treatment (sensitization).

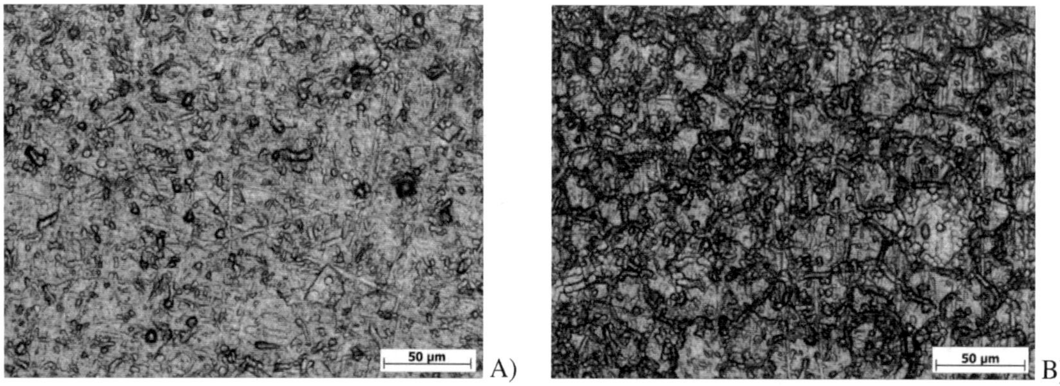

Figure 4. Microstructure of AISI 316L after the etch test: A) "as received" state; B) the state after heat treatment (sensitization).

As shown in figures 3B and 4B the microstructures of 304 and 316L steels after sensitization can be classified as the ditch ones. Carbides, precipitated at grain boundaries during the heat treatment, dissolved during the etching process and ditches were formed. Ditches in AISI 316L microstructure (figure 4B) are lighter-colored than the ones in the AISI 304 specimen (figure 3B). A lighter color of the etched ditches indicates their smaller depth compared to the dark ones. Accordingly, AISI 304 steel after sensitization exhibited higher susceptibility to the intergranular corrosion than AISI 316L steel. Also the authors [7] observed similar ditch structures of austenitic steels.

Unlike 304 and 316L steels, AISI 316Ti steel stabilized by titanium reflected the minimal difference between the "as received" and the sensitized state (figure 5), without the grains surrounded by ditches. According to performed etch test it points to the highest resistance to the intergranular corrosion.

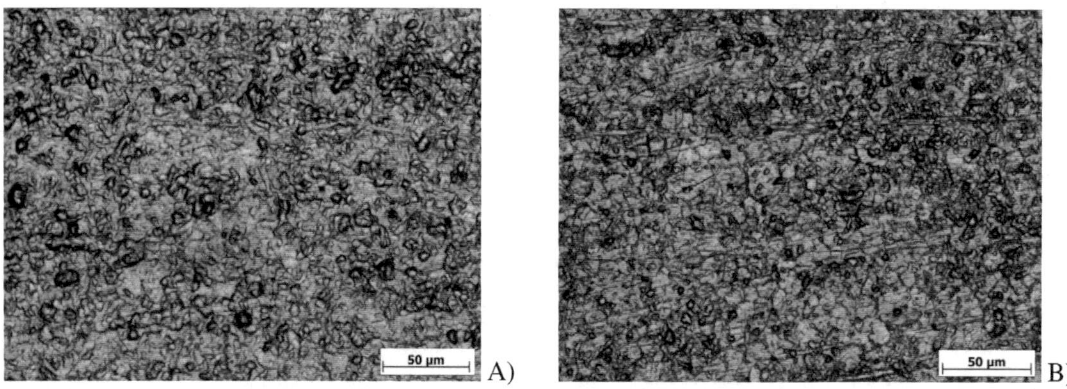

Figure 5. Microstructure of AISI 316Ti after the etch test: A) "as received" state; B) the state after heat treatment (sensitization).

Strauss test (E practice of ASTM A262 standard method) is usually used for a confirmation of the susceptibility to the intergranular corrosion in the case of the positivity of the etch oxalic acid test (appearance of ditch structures). In this study the Strauss test was carried out on all specimens regardless of results of the etch test. The aim was to compare the both tests and to find out witch one is more precise, comfortable and which one suits to the quick assessment the susceptibility of austenitic steels to the intergranular attack. The microstructures after Strauss test are shown in figures 6–8.

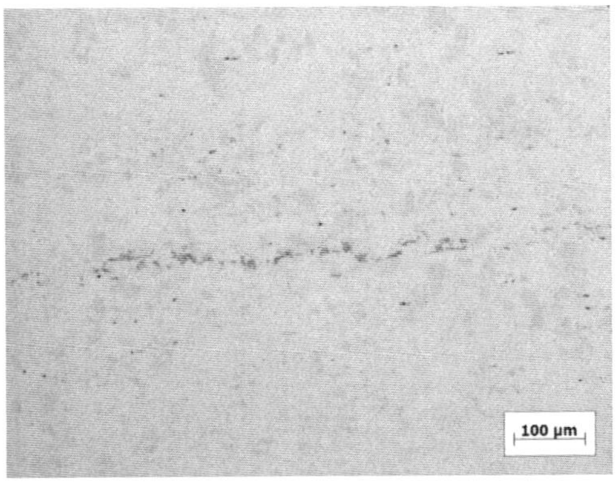

A)

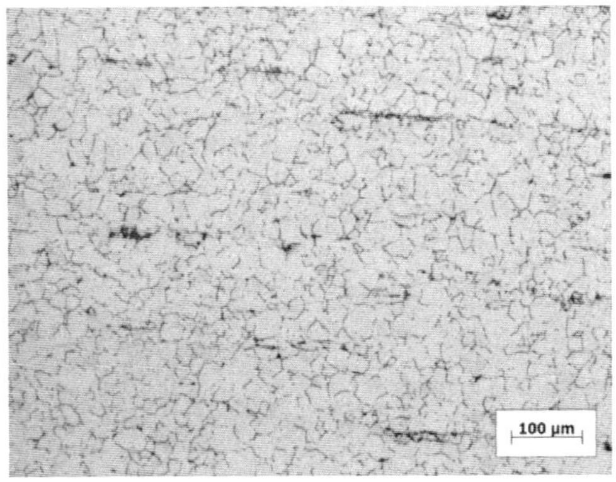

B)

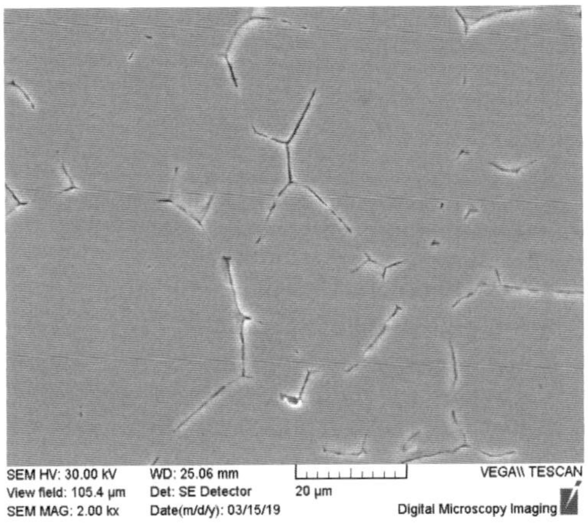

C)

Figure 6. Microstructure of AISI 304 after Strauss test: A) the "as received" state; B) the state after heat treatment (sensitization); C) the state after heat treatment (sensitization) – detail of the attacked grain boundaries (SEM).

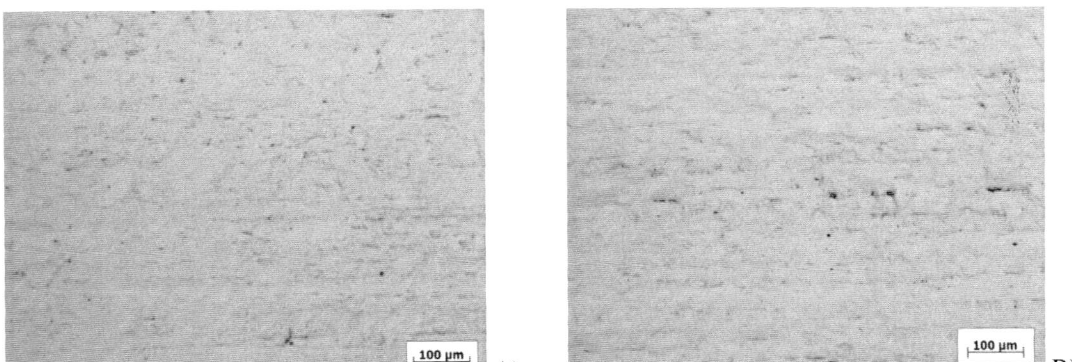

Figure 7. Microstructure of AISI 316L after Strauss test: A) "as received" state; B) the state after heat treatment (sensitization).

Strauss test confirmed a susceptibility of AISI 304 stainless steel exposed to the critical temperatures to the intergranular corrosion (figure 6B, 6C). 15-hour exposition of the sensitized specimen in aggressive acidic hot solution evoked marked corrosion attack at the chromium depleted zone along the grain boundaries. As can be seen in figure 7B, unlike the etch test AISI 316L influenced by the critical temperatures did not reflect any corrosion attack at the grain boundaries. This result points to the relevance of the both independent tests in the case of light colored (= shallower) ditches at the grain boundaries obtained by the etch test. Both performed tests confirmed the high resistance of AISI 316Ti stainless steel to the intergranular corrosion (figure 8).

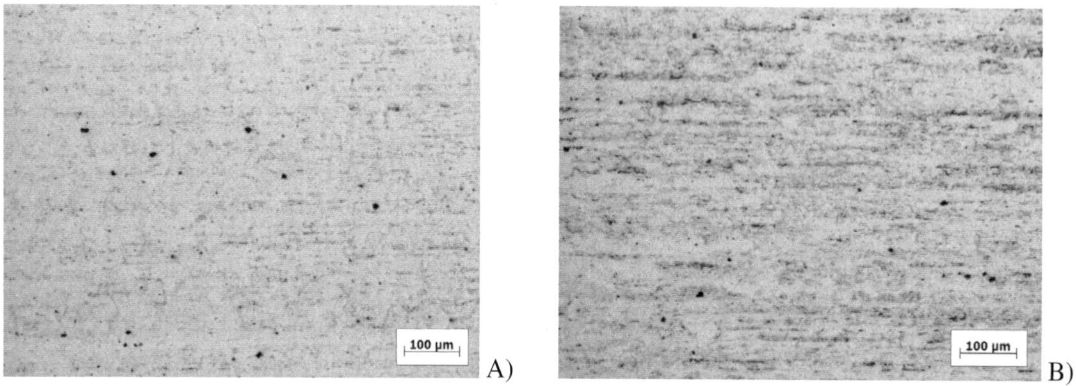

Figure 8. Microstructure of AISI 316Ti after Strauss test: A) “as received” state; B) the state after heat treatment (sensitization).

5. Conclusion

Based on the results of performed experiments, it can be concluded:

- Microstructures after heat treatment (sensitization) visualized by classical chemical etching (Kallings 2) showed no radical changes compared to the “as received” state and a precipitation of the chromium carbides at the grain boundaries was not observed.
- The microstructures of the tested steels after the etch test (A practice) in the "as received" state did not show ditch structures caused by the dissolution of chromium carbides at the grain boundaries. After improper heat treatment (sensitization), the microstructures of AISI 304 and AISI 316L steels after etch test showed a ditch attack at the grain boundaries. In the AISI 316Ti microstructure the similar ditch structures were not observed.
- Strauss test (E practice) confirmed high susceptibility of improperly heat treated AISI 304 and high resistance of AISI 316Ti stainless steels to the intergranular corrosion. Strauss test did not confirm results of the etch test of the sensitized AISI 316L, the intact grain boundaries indicated high resistance to the intergranular corrosion.
- All three steels in “as received” state showed high resistance to intergranular corrosion and it means they were not exposed to critical temperatures during the production process (they were not sensitized).

According to the performed experiments the etch test can be considered more comfortable than Strauss test. Unlike Strauss test the microstructures obtained by the etch test were well readable even without metallographic preparation of the specimens. Except for AISI 316L specimen (ditch structure with light coloured shallow ditches) the etch test brought clear results.

Acknowledgement

The research was supported partially by Scientific Grant Agency of Ministry of Education, Science and Sport of Slovak Republic and Slovak Academy of Sciences grants KEGA No. 049ŽU-4/2017 and No. 013ŽU-4/2019.

This work was also supported by the project of Operational Programme Research and Innovation: Research and development activities of the University of Žilina in the Industry of 21st century in the field of materials and nanotechnologies, No. 313011T426. The project is co-funding by European Regional Development Fund.

References

[1] Tukur S A, Dambatta M S, Ahmed A and Mu'az N M 2014 Effect of Heat Treatment Temperature on Mechanical Properties of the AISI 304 Stainless Steel *Int. J. Innov. Res. Sci. Eng. Technol.* **3** (2) 9516-9520

[2] Abd Rashid M W, Gakim M, Rosli Z M and Azam M A 2012 Formation of $Cr_{23}C_6$ during the Sensitization of AISI 304 Stainless Steel and its Effect to Pitting Corrosion *Int. J. Electrochem. Sci.* **7** 9465-9477

[3] Bansod A V, Patil A P, Moon A P and Khobragade N N 2016 Intergranular Corrosion Behavior of Low-Nickel and 304 austenitic Stainless Steels. *JMEPEG* **25** 3615–3626

[4] Kumar V, Joshi P, Dhakad S, Shekhar H, Singh S and Kumar S 2015 Analysis of the Effect of Sensitization on Austenitic Stainless Steel 304L Welded by GTAW Process *HCTL Open Int. J. Technol. Innov. Research (IJTIR)* **14** http://ijtir.hctl.org

[5] McGuire M F 2008 *Stainless Steels for Design Engineers. s. l.* (ASM International)

[6] Khatak H S and Raj B 2002 *Corrosion of Austenitic Stainless Steels, Mechanism, Mitigation and Monitoring* (ASM International)

[7] Dománková M, Marek P and Moravčík R. 2007 The effect of cold work on the sensitisation of austenitic stainless steels *Mater. Tehnolog.* **41** (3) 131

[8] Oršulová T, Palček P, Roszak, M, Uhríčik M and Kúdelčík J 2018 Change of magnetic properties in austenitic stainless steels due to plastic deformation *Procedia Structural Integrity* **13** 1689-1694

[9] Jambor M, Nový F, Bokůvka O, Trško L and Oravcová M 2018 Influence of structure sensitising of the AlSi 316Ti austenitic stainless steel on the ultra-high cycle fatigue properties *MATEC Web of Conferences* **157** 05011

[10] *Available at:* www.italinox.sk (19.6.2019)

[11] ASTM A262 *Standard Practices for Detecting Susceptibility to Intergranular Attack in Austenitic Stainless Steels* (ASM International)

IOP Conf. Series: Materials Science and Engineering **726** (2020) 012018 doi:10.1088/1757-899X/726/1/012018

Investigations of microstructure and mechanical properties of brass alloys produced by sand casting method at different casting temperatures

H Yilmaz Atay, G Uslu, Y Kahmaz and Ö Atay

İzmir Katip Çelebi University, Faculty of Engineering and Architecture, Department of Materials Science and Engineering, Çiğli 35620 İzmir, Turkey

hgulyilmaz@gmail.com

Abstract. Brass, which is an alloy of copper and zinc, is of historical and lasting importance due to its hardness and workability. Indeed, to make this metal alloy, ancient metalworkers around the Mediterranean Sea were able to melt copper with zinc as early as 3000 B.C. Brass is stronger and harder than copper. The amount of copper varies between 55% and 95% by weight, depending on the type of brass and its intended use. It is easy to form into various shapes, it is a good conductor of heat and generally resistant to corrosion from salt water. It can be used to make pipes and tubes, weather-stripping and other architectural trim pieces, screws, radiators, musical instruments, and cartridge casings for firearms. The manufacturing process involves combining the appropriate raw materials into a molten metal, which is allowed to solidify. The shape and properties of the solidified metal are altered through series of carefully controlled operations to produce the desired brass alloy. Temperature used in the process is one of the parameter to affect the product properties. In this study, the brass alloys were produced by a sand casting method at different casting temperatures. Structural and mechanical properties were observed. Optical microscope, XRD and the SEM analysis were performed for the microstructure and phase analysis of the material. Charpy impact test and tensile test were applied to consider the mechanical properties of the samples. As a main purpose of this study, the results were evaluated and then the optimum temperature for the brass material production was determined.

1. Introduction

Brass is the most important alloy of copper. Because of its yellow colour, the copper-zinc alloy is known as “yellow”. It has easy workability, high corrosion resistance and beautiful appearance. In addition, brass has higher toughness than bronze which is a copper-tin alloy. History of the brass dates too far back. The pre-dynastic Egyptians knew copper very well, as proved in hieroglyphs. The Greeks described brass as 'orefalos'. Some Roman writers mentioned brass in their work and called it 'Aurichalum'. It was used for production of coins and golden helmets. They also preferred brass for all kinds of jewelry and for ornamented works. Actually, after the Copper Age, the Bronze Age came, and was later followed by the Iron Age. There was no 'Brass Age', because it was not easy to make brass [1-5].

Brass contains other metals in its structure according to its intended use. Addition of 10% zinc to copper causes the bronze colour, while 15% adds gold colour and 20-38 % yellow colour. When the amount of zinc added achieves 45%, a silvery white colour is obtained. The alloy exhibits a brittle structure [1, 2]. Brass has higher malleability than copper. The low melting point of brass is 900 to 940 °C, and it is easy to cast. The density of brass is 8.4 to 8.73 grams per cubic centimeter (0.303 to 0.315 lb/cu in) [3].

Brass used for commercial purposes is separated into the alpha and alpha-beta brass, according to their structure. Alpha brass is a kind of brass that contains only an alpha phase. Unbalanced cooling causes that beta phase is formed in very small amounts. Owing to their ductility at room temperature, they can be easily processed at cold temperatures. If a high amount of cold deformation is applied to the alpha brass, annealing and slow cooling must be applied to remove the stress. In this case, the brass should be annealed to a temperature of 625 °C. In case of higher temperatures, grain growth occurs. It can be easily applied to alpha brass, such as deep drawing, bending and cold rolling. Alpha-beta brass contains 54-61 % of copper. The structures have hard and brittle α and β' phases at the room temperature. When high temperatures are reached, β phase is formed. For this reason, plastic forming to $\alpha + \beta$ phases of brass are applied at high temperature [3]. As it affects the properties of the brass, production process is important. Especially with the development of the industrial revolution, brass production has diversified. In 1738, William Champion succeeded in obtaining a patent for the production of zinc through the distillation of calamine and charcoal. Although the first rolling mills were established in the 17th century, the establishment of strong rolling mills came in the middle of the 19th century [4-7]. Muntz made it possible to make cheap, hot-processed brass plates with 60/40 brass in 1832 [4]. In 1957, Bungardt published a patent in which he selected three different casting temperatures and investigated the effect of temperature [8].

In this project, microstructure and mechanical properties of brass materials produced in a sand mould at different temperatures were observed and evaluated. Sand casting, also known as sand moulded casting, is a metal casting process characterized by using sand as the mould material. During this process, material is heated to the correct temperature to melt. Sometimes, it is treated to modify the chemical composition, in order to achieve the required material properties. Then the molten metal is poured into a mould of the desired shape cavity to cool down and solidify [9]. Design flexibility, high complexity shapes, wider material choice, low cost tooling and short lead time are the advantages of this method. After obtaining the samples, Optical microscope, XRD and SEM-EDS analyses were performed for the microstructure and phase analysis of the material. Also, the tensile-bending strength and elongation amounts of the mechanical properties of the material were tested.

2. Materials and methods

The raw materials used in experimental studies and provided by EMSA DÖKÜM were the scrap materials from other companies' processes (Figure 1). The current paper is a recycling study, in this respect.

Figure 1. Raw materials and a furnace.

IOP Conf. Series: Materials Science and Engineering **726** (2020) 012018 doi:10.1088/1757-899X/726/1/012018

The brass alloy production started by melting the scrap. The obtained melt was poured into sand moulds which were prepared separately for each group. Casting processes were applied in three different temperatures of 1000 °C, 1100 °C and 1200 °C. The brass production for this purpose consisted of five stages. The steps are given in the flowchart diagram in Figure 2.

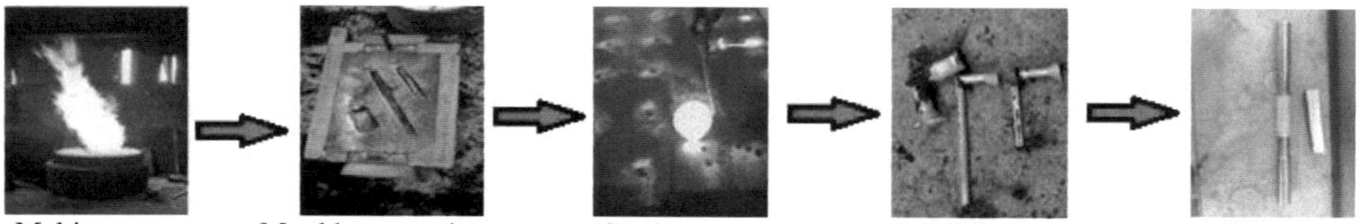

Figure 2. Flowchart of brass production process.

The moulding process is done using mould sand. It consists of silica, clay, water and other materials. The gap of the model is formed in the compressed sand. The melted alloy is poured into the mould cavity and allowed to solidify (Figure 3). The surface separator powder is poured on the surface of the model. The sand mass around the model is carefully compacted.

Figure 3. Moulding steps.

The sum of the ducts that allow the melt to reach the mould cavity is called a moulding runner system. The material melted from the pot passes through the horizontal and vertical channels forming the gating system, and fills the mould cavity. The ducts that are opened ensure that the gases emerge from the mold during the casting. When the mould is filled with the liquid metal, the mould gases are trapped in the gradually decreasing mould cavity. They form a pressure within the mould. These gases may cause various casting faults. The liquid metals and alloys pouring into the mould cool down and solidify, while their volumes shrink. The resulting gaps are called collapses. Collapses cause faulty castings. The feeders move the debris out of the part.

After casting, the samples were allowed to solidify. It is not possible to solidify all of the liquid metal filled into the mould at the same time. Solidification first begins in thin sections where cooling is rapid. Spontaneous shrinkage and volume reduction occur in these regions. Then these regions are fed with the material which is still liquid in the thick regions. After the solidification and cooling of the brass, the casting is separated from the sand mould.

Regarding the mould type, a two-piece model was used in the process. The bottom and top moulds are demonstrated in Figure 4.

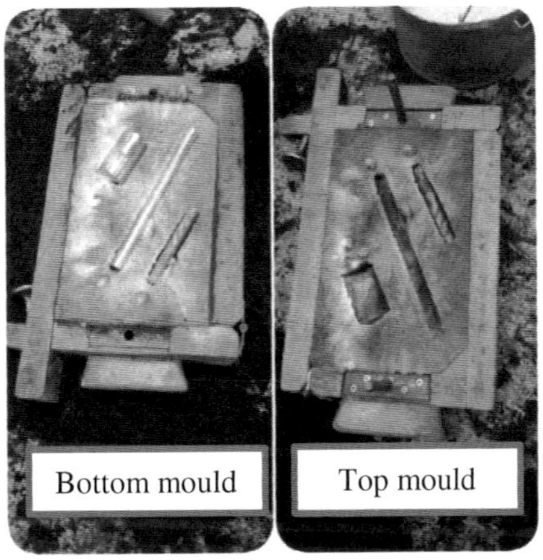

Figure 4. Two-piece model of a mould.

After obtaining the cast samples, the desired dimensions were provided by using the milling and cutting equipment. The samples were processed on CNC machines to be prepared for mechanical research. The phase analysis of the metal alloys was performed with the help of a Empyrean X-ray diffractometer with Cu Kα radiation (λ=1.540 Å), in the 2θ range of 20-800. The surface morphology was characterized by using a Scanning Electron Microscope (SEM, Zeiss Sigma 300 VP-FESEM). Mechanical behavior of the samples was tested by the Charpy impact test machine (ALŞA) and a tensile testing machine (SHIMADZU AGS-X) according to the EN ISO 148-1.

3. Results and discussion

XRD analysis results of the obtained samples are demonstrated in Figure 5. The equivalent of the peaks seen in XRD test analysis is equivalent to Pd1.15Cu2.85 for Sample 1, Sb0.20Cu3.80 for Sample 2, and Zn0.20Cu3.80 for Sample 3.

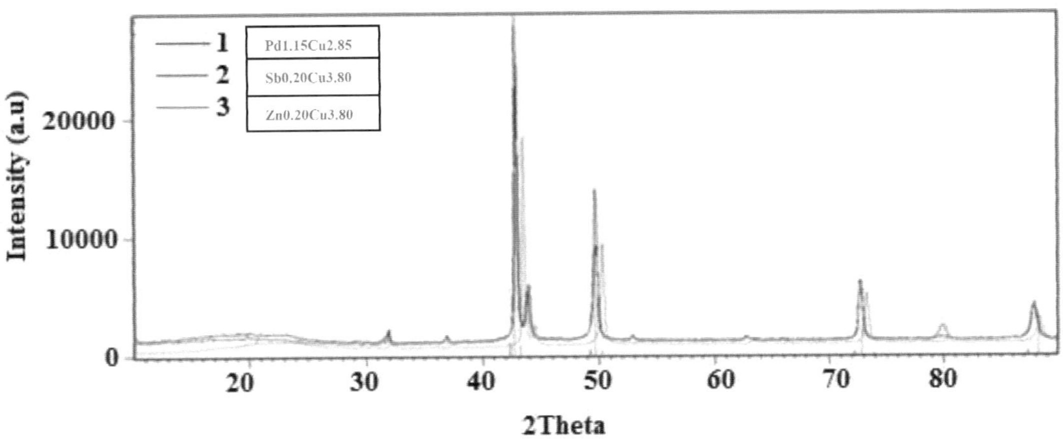

Figure 5. XRD Graphic of Count -2 Theta.

An X-ray fluorescence spectrometer was used to determine the elemental and chemical composition of our alloy. The results are shown in Table 1. According to our XRF test results, the concentration of copper was 58.84 %, and the concentration of zinc was 32.83 %.

Table 1. XRF test results.

Symbol	Element	Concentration
Mg	Magnesium	0.28 %
Al	Aluminum	0.25 %
P	Phosphorus	0.15 %
Fe	Iron	0.38 %
Ni	Nickel	0.25 %
Cu	Copper	58.84 %
Zn	Zinc	32.83 %
Sn	Tin	0.32 %
W	Tungsten	0.58 %
Pb	Lead	2.42 %

The optical microscope was used to observe the effect of the casting temperature on the microstructure. Samples were cut, grinded, polished and etched in order to examine the samples under optical microscope. Etching was carried out with ethanol. Figure 6 depicts the optical analysis results of the samples. As the casting temperature fell from 1200 °C to 1000 °C, a smaller particle size was observed in the third sample (1000 °C). One of the most important problems encountered in casting process are pores. In the microstructures of the samples, homogenous pores were observed as black spots. A smaller particle size was observed in the microstructure of our third sample having 1000 °C casting temperature. A more homogeneous structure was seen. In the first sample having 1200 °C casting temperature, irregular and larger grain size was observed.

When casting at different temperatures, the low temperature material will cool down earlier. As mentioned above, the complete cooling of the poured material cannot be achieved in an instant. The earlier cooling zones are fed with liquid material from the later cooling zones due to the shrinkage and volume reduction. When this liquid transfer is completed, then grains are formed. If the solidification occurs before this transfer is completed at low temperature, a network image is obtained, as shown in Figure 6.

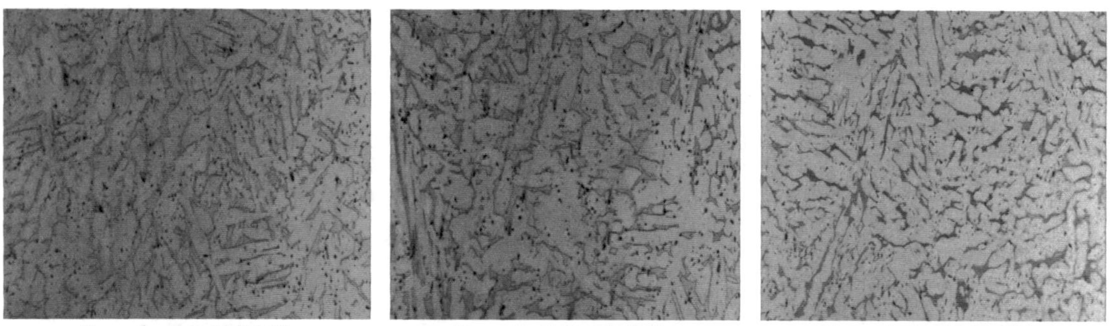
Sample 1 (1200 °C) Sample 2 (1100 °C) Sample 3 (1000 °C)

Figure 6. Microstructure of prepared samples (magnification 100 x).

The results of Charpy impact test are shown in Table 2. This test is the process of determining the energy used to break the sample with a hammer at the tip of a pendulum at the end of the sample resting on the 2-notch base in horizontal and simple beam and the effect of multi-axis stresses occurring at the base of the notch. As can be seen, the average fracture energy of our first sample (casted at the casting temperature of 1200 °C) is according to the impact test 2.33 kpm. The second sample with the casting temperature of 1100 °C has the average fracture energy of 2.57 kpm and the third sample with the casting temperature of 1000 °C 2.63 kpm. According to the Charpy impact test results, the third sample having 1000 °C casting temperature had higher breaking energy compared to samples casted at 1200 °C and 1100 °C. This finding in the current study is similar to Jha's research [10] proving that fracture energy had increased by decreasing the temperature.

Table 2. Charpy test results.

Sample No	Wall thickness (mm)	Width (mm)	Test temperature (°C)	First energy (kpm)	Last energy (kpm)	Fracture energy (kpm)	Average fracture energy (kpm)	Measurement error
1 (1200 °C)	10.00	8.10	Avg. 19 °C	0	2.30	2.30	2.33	2.33 ± 0.058
	9.96	8.12	"	0	2.30	2.30		
	9.96	8.12	"	0	2.40	2.40		
2 (1100 °C)	9.94	7.88	Avg. 19 °C	0	2.60	2.60	2.57	2.57 ± 0.058
	9.96	8.00	"	0	2.60	2.60		
	9.90	7.87	"	0	2.50	2.50		
3 (1000 °C)	9.96	8.04	Avg. 19 °C	0	2.60	2.60	2.63	2.63 ± 0.058
	10.00	8.12	"	0	2.70	2.70		
	9.96	8.08	"	0	2.60	2.60		

The tensile test results are depicted in Figure 7. It is used to determine the behaviour of a sample when applying an axial tensile load. Tensile test was applied to the samples at a speed of 5 mm/min.

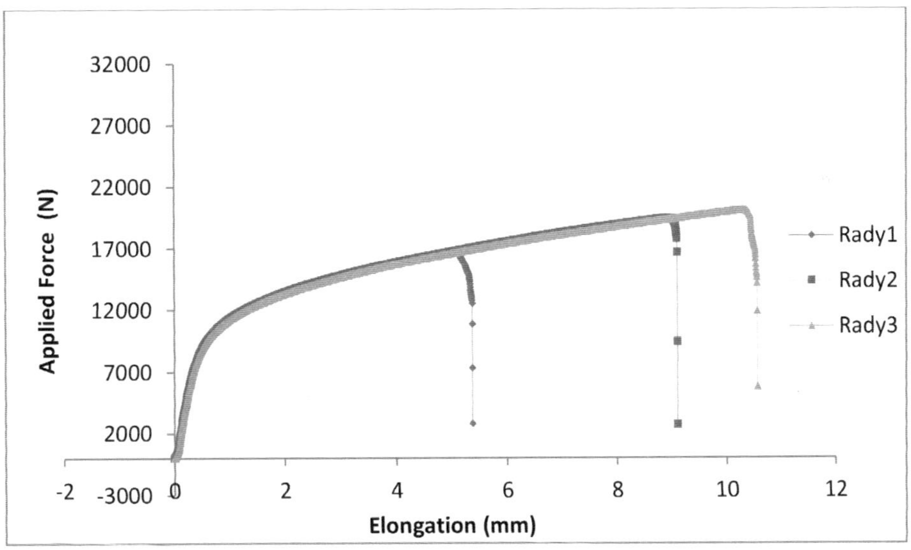

Figure 7. Tensile test results of the samples.
(Seri 1=Sample 1, Seri 2=Sample 2, Seri 3=Sample 3)

Table 3 demonstrates the total tensile test results. According to those results, the maximum stresses are 212 N/mm^2, 247 N/mm^2 and 255 N/mm^2. Therefore, decreasing the temperature increased the stress value. Regarding the elongation, it increased from 5.06 mm to 10.29 mm by decreasing the temperature. It can be said that ductility also improved in this way. The obtained results are similar to Ref [10].

Table 3. Results of tensile tests.

	Elongation (mm)	Maximum stress (N/mm^2)
Sample 1 (1200 °C)	5.06	212
Sample 2 (1100 °C)	8.92	247
Sample 3 (1000 °C)	10.29	255

SEM analysis results are demonstrated in Figure 8. As a result of this analysis, the fraction patterns of the first sample (1200 °C) and the second sample (1100 °C) show a relatively brittle fracture. However, it can be said that there is higher ductile fraction in the third sample (1000 °C). In particular, the full depression obtained in the third image reminds us of the concave shape of ductile fracture. It did not occurre in brittle fractures in the inner parts of surfaces. The results are similar to Ref [11]'s results in the macroscopic or binocular microscope observations; the fracture surfaces of the Charpy specimen have similar patterns.

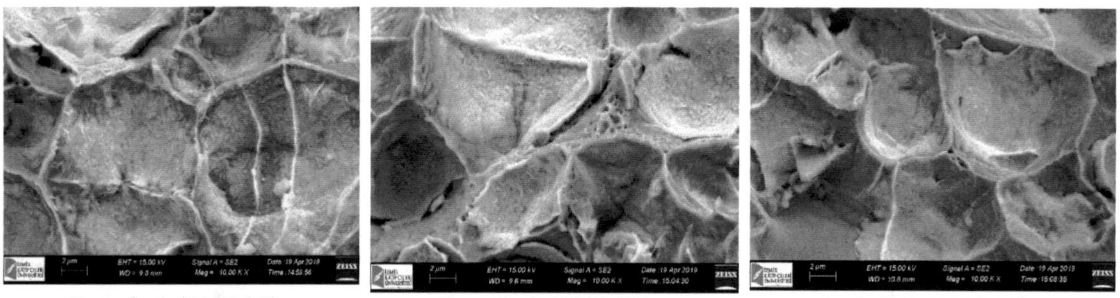

Sample 1 (1200 °C) Sample 2 (1100 °C) Sample 3 (1000 °C)

Figure 8. SEM images of the samples.

All these results were evaluated, and the optimum temperature for the brass material was determined. It is our greatest ambitions that the final product, mechanical properties of which were tested and developed, is brought into our present life.

4. Conclusion

In this study, the effect of casting temperature on the mechanical properties of brass alloy was investigated. Mechanical properties were tested, all results were evaluated and the optimum temperature for the brass material was determined. In the third sample (1000 °C), the temperature was observed to improve the mechanical properties. According to the Charpy and tensile test results, the third sample (1000 °C) exhibited more ductile behaviour under higher load, and the elongation value was higher than that of the second (1100 °C) and the first (1200 °C) samples. Tensile tests and the Charpy test were supported by SEM, XRD and optical microscope. At the end of the analyses, the optimum temperature was found to be 1000 °C. At this temperature, mechanical properties were improved and energy saving was obtained. The cost of production were positively influenced.

References

[1] Brady G S 1991 *Materials Handbook: an encyclopedia for purchasing managers engineers, executives and foremen* (New York: McGraw-Hill Book Company Inc.)

[2] OSH Answers: *Non-sparking tools*. Ccohs.ca (2011-06-02). Retrieved on 2011-12-09.

[3] Walker R "Mass, Weight, Density or Specific Gravity of Different Metals". *Density of Materials*. United Kingdom: SImetric.co.uk. Retrieved 2009-01-09. 8430–8730.

[4] *The Copper Age. Available at:* <https://copperalliance.org.uk/about-copper/copper-history/copper-through-the-ages/> Accessed on 27.04.2019.

IOP Conf. Series: Materials Science and Engineering **726** (2020) 012018 doi:10.1088/1757-899X/726/1/012018

[5] Haque M and Khan M. 2008 Investigation on Structure and Properties of Brass Casting. *J. Mater. Sci. Technol.* **24** (3) 299-301
[6] Schiffer P, Schiffer N and Schiffer H 1977 *The Brass Book* (Schiffer Pub Ltd) ISBN 0-916838-17-X.
[7] Igelegbai E, Edward A, Oluwaseun A, Adefemi A and Ilesanmi D 2016 Evaluation of Mechanical and Microstructural Properties of α-Brass Alloy Produced from Scrap Copper and Zinc Metal through Sand Casting Process *J. Miner. Mat. Char. Eng.* **5** 18-28
[8] Bungardt W 1957 Process For Manufacturing Brass and Bronze Alloys Containing Lead, US2802733A, US
[9] *Available at:* <https://engineeringproductdesign.com/advantages-disadvantages-sand-casting/> . Accessed on 27.04.2019.
[10] Sujit J, Devibala B and Rajalingam P 2015 Experimental analysis of microstructure and mechanical properties of copper and brass based alloys *IJAME-Int. J. Automotive and Mechanical Engineering* **11** 2317-2331
[11] Abid D, Ktari A, Mellouli D, Gafsi N and Haddar N 2019. Failure Analysis of Shot – Sleeves Used In Brass High Pressure Die – Casting Process. *Eng. Fail. Anal.* **104** 177-188

IOP Conf. Series: Materials Science and Engineering **726** (2020) 012019 doi:10.1088/1757-899X/726/1/012019

Structural characteristics of NiTi alloys after thermal treatment

J Drápala[1], M Losertová[1], M Štencek[1], K Konečná[1], D Matýsek[2] and M A Sevostyanov[3]

[1]VŠB – Technical University of Ostrava, Faculty of Materials Science and Technology, 17. listopadu 15/2172, 708 00 Ostrava - Poruba, Czech Republic
[2]VŠB – Technical University of Ostrava, Faculty of Mining and Geology, 17. listopadu 15/2172, 708 00 Ostrava - Poruba, Czech Republic
[3]Baikov Institute of Metallurgy and Material Sciences, RAS, Leninskii pr. 49, Moscow, Russian Federation

mlosertova@vsb.cz

Abstract. The study of the effect of thermal treatment conditions on NiTi alloy used in biomedicine was carried out. The experimental samples of NiTi alloy with 55.8 wt% Ni were subjected to the heat treatment. The thermal regimes consisted of aging treatment at 300, 350 and 400 °C for 30 minutes followed by air cooling. The microstructures before and after the heat treatment were observed by optical microscopy. The study was completed by the microhardness measurement and XRD analysis. The obtained results show that the aging temperatures led to microstructural and microhardness changes for the investigated alloy.

1. Introduction

Shape memory alloys based on NiTi, commonly known as "Nitinols", are nowadays successfully used in different areas of the medicine, as for instance plates and nails in traumatology, vascular stents, surgery catheters, root canal instruments or orthodontic braces. Other properties, in addition to the shape memory behaviour, are also very good corrosion resistance, biocompatibility, advantageous strength/density ratio or high damping capacity.

The shape memory effects and superelastic behavior of Ni-rich NiTi alloys, it means more than 50.6 at% Ni, are significantly dependent on microstructural modifications provided through the heat treatment [1-6]. It is generally known that thermal treatment of Ni-rich NiTi alloys induces diffusion processes that lead to precipitation of different phases: metastable Ni_4Ti_3, metastable Ni_3Ti_2 or stable Ni_3Ti [3-5]. The microstructural changes have a major impact on the austenite-martensite transformation temperatures and consequently on the martensitic transformation in NiTi alloys. Owing to an appropriate thermal regime the alloy behavior can be modified according to requirements [7]. Moreover, the precipitation of secondary phases can also contribute to a substantial increase in the strength of NiTi alloys. Farther critical factor influencing both transformation and microstructure characteristics of the NiTi material is the alloy composition.

The work is focused on the effect of heat treatment on NiTi alloy containing 55.8 wt% Ni (50.7 at% Ni). The microstructure of experimental specimens was observed by optical microscopy in the condition of as-received and heat treated. The microhardness measurement and X-Ray diffraction (XRD) analysis appeared as useful tool for the study of microstructure changes.

2. Experimental

Commercially available NiTi wire composed of 55.8 wt% Ni, ≤ 0.05 wt% O, ≤ 0.02 wt% C and balance of Ti (SE508 alloy) were subjected to heat treatment (aging). The LAC chamber furnace with resistance heating was used for annealing. The thermal regimes that consisted of holding at 300, 350 and 400 °C for 30 minutes followed by air cooling are shown in Figure 1.

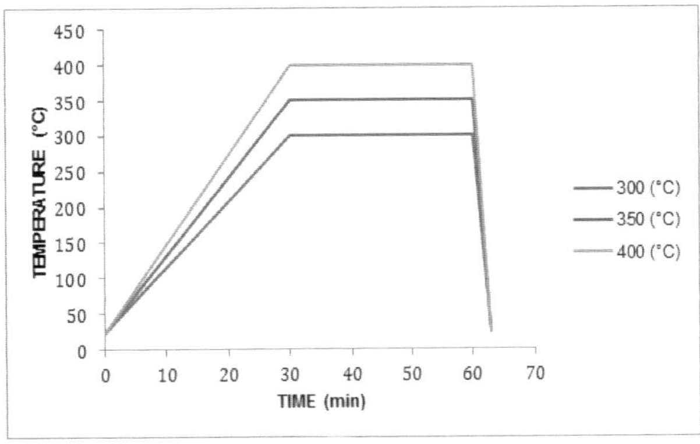

Figure 1. Schematic of thermal treatment regimes applied for NiTi specimens.

For the microstructure study longitudinal cuts of the samples were cold mounted into LevoCit (Struers) resin, ground, polished and etched. The martensite microstructure was revealed by immersing into anetching solution of CH_3COOH:HF:HNO_3 (in ratio of 3:3:5) for 4–5 s. The microstructure features before (as-received) and after the aging were observed by optical microscopy (OLYMPUS GX51). The study was completed by Vickers microhardness measurement using FUTURE-TECH FM-100 automatic microhardness tester with FM-ARS900 control unit. XRD analysis was performed by means of Bruker-AXS D8 device equipped with Lynx-Eye PSD (CuKa source/Ni filter). As the samples were observed only in longitudinal sections, the microhardness was measured using five impacts performed in the central area parallel to the edge of the samples and by loading of 0.1 kg. The X-Ray diffraction was measured under following conditions: accelerating voltage 40 kV, anode current 40 mA, step mode – step 0.0147, range 5–80°, detailed measurement of the spectrum in the range from 38–47°, where appropriate conditions exist.

3. Results and discussion

The microstructure observed for all samples is given in Figure 2 and is formed of remaining austenite and needle/plate-like martensite with different space distribution and orientation, thickness or length. The abundance of both phases seems to be depending on the aging temperature. For the samples in as-received and at 400 °C heat treated conditions, it is possible to observe a microstructure formed by coarse martensitic needles and plates with remaining austenite (Figure 2a and 2d). Conversely, very fine martensitic needles that seem in principle to constitute all surface observed are seen for the sample after the aging at 350 °C (Figure 2c). The highest degree of arrangement of martensitic needles to particular packets is revealed in the sample heat treated at 300 °C.

The particles $NiTi_2$ of various sharp-edged shapes, fractured and aligned into parallel lines, as seen in Figure 2, are present in the samples as consequence of wire processing technology. The length of the particles alignment achieves up to several tens of micrometers in some cases. Moreover, the presence of $NiTi_2$ phase is observed in the microstructure of all samples, regardless of thermal treatment conditions, as it was confirmed in our previous SEM observation and EDX analysis [8]. As the NiTi wires serve as starting material for endodontic instruments producing the existence of sharp-edged $NiTi_2$ particles can be generally considered as negative for the mechanical behavior of the material. In practical use, NiTi instruments can be damaged by fatigue cracks initiated at the particle edges when subjected to significant rotation stresses.

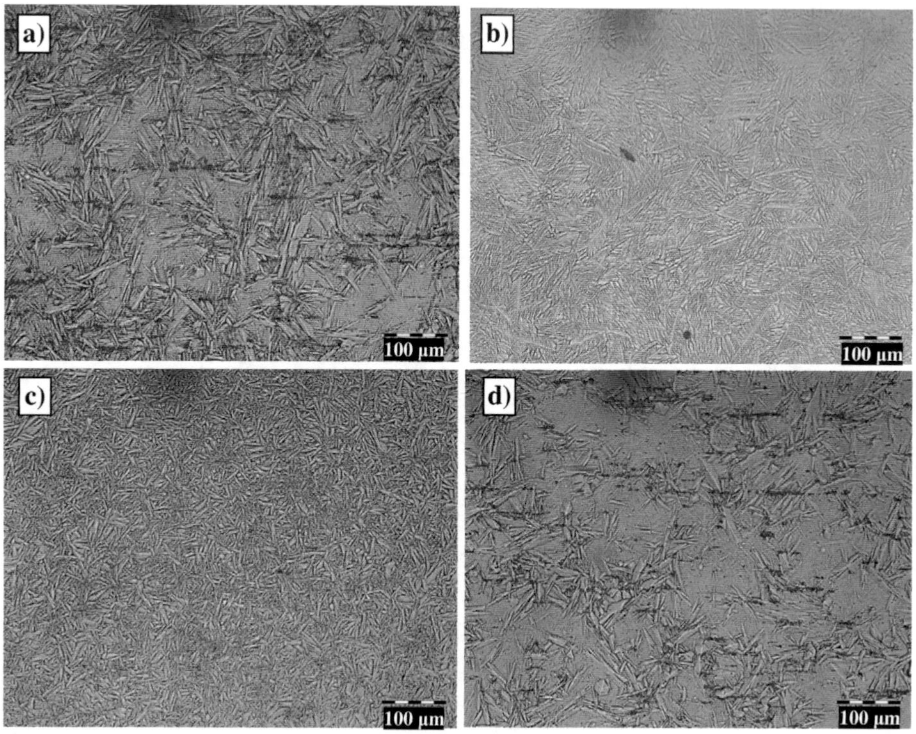

Figure 2. Microstructure of NiTi wires in the conditions: a) as-received and heat treated, b) at 300 °C, c) 350 °C and d) 400 °C.

The average values of microhardness HV0.1 that are summarized in Table 1 range from 373 to 406. The changes in the microhardness due to the heat treatment are very small but can be related with the martensite morphology and distribution. The highest average value of 406 HV0.1 was measured for the sample heat treated at 350 °C (HT350) and corresponds to the very fine but not especially arranged martensitic needles (Figure 2c). The lowest microhardness (373 HV0.1) for the sample heat treated at 400 °C (HT400) is related with the different character of the microstructure composed of coarse martensite needles and laths. Nevertheless, the sample aged at 300 °C with highly arranged martensite needles reaches of 390 HV0.1 that is close to the value of the as-received stage but with quite different microstructure morphology.

Table 1. Microhardness HV0.1 of NiTi samples in conditions: as-received, heat treated at 300 °C (HT300), 350 °C (HT350) and 400 °C (HT400).

Specimen	HV0.1
As-received	395 ± 47
HT300	390 ± 14
HT350	406 ± 29
HT400	373 ± 29

Unlike microstructures, the XRD patterns are very similar for the samples in the conditions as-received and heat treated at 300 and 350 °C (Figure 3a, 3b and 3c). In the XRD charts two principal peaks corresponding for austenite phase (parent B2 structure of NiTi) and R-phase NiTi (rhombohedral distortion of B2 phase) were only detected.

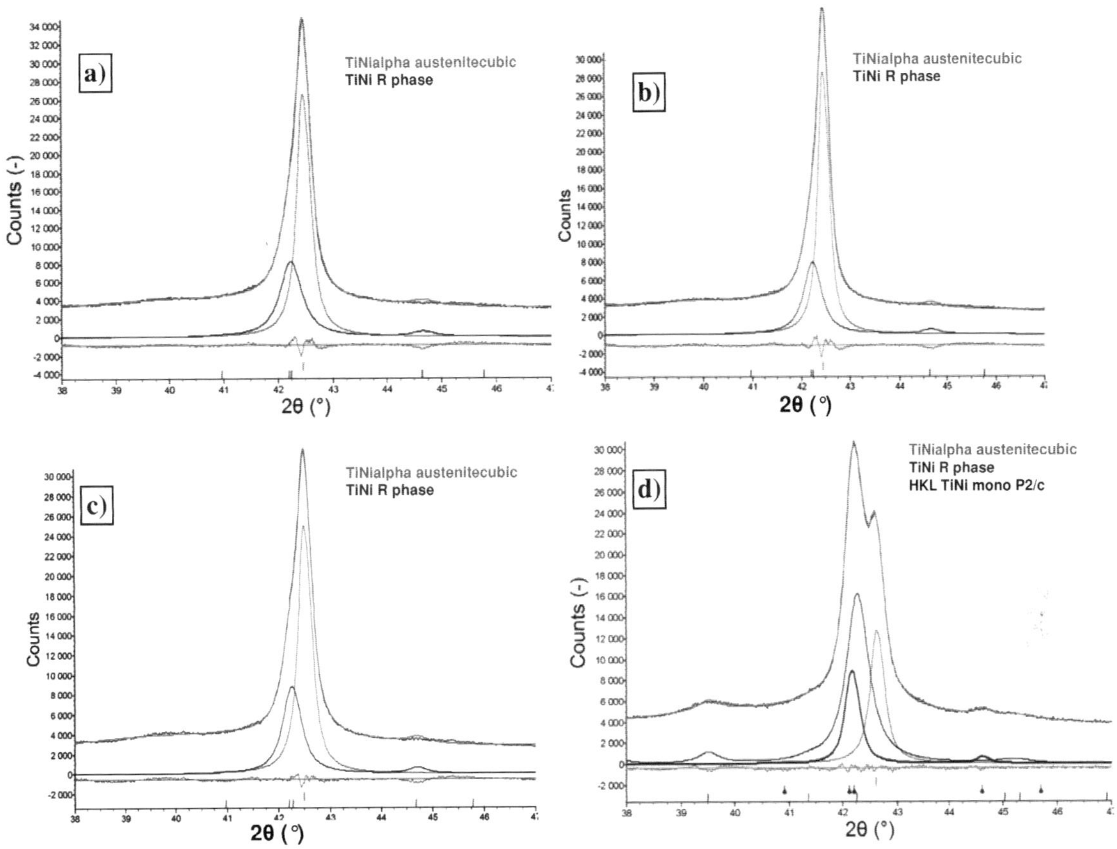

Figure 3. XRD patterns of the NiTi samples in the conditions: a) as received; heat treated at b) 300 °C, c) 350 °C and d) 400 °C.

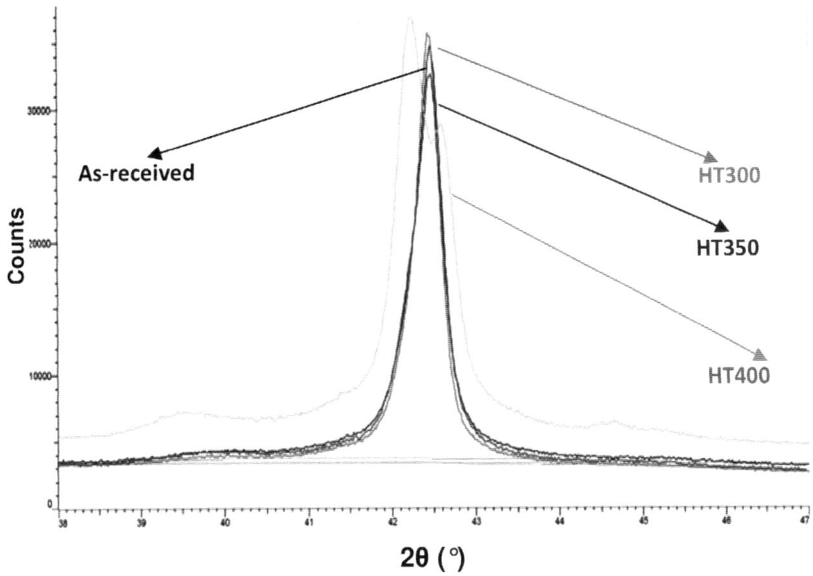

Figure 4. Comparison of XRD patterns for the samples in four heat treatment conditions.

The evolution of the XRD chart of the sample aged at 400 °C shows differences in the presence of three spectra, where the monoclinic martensitic phase is appearing together with B2 and R-phase. The formation of monoclinic martensite (B19') corresponds to the lowest microhardness measured (Table 1). Due to inhomogeneous distribution of $NiTi_2$ phase and most probably nanometric sizes of metastable Ni_4Ti_3 or Ni_3Ti_2 particles, no other evident peaks than that ones for austenite and martensite phases were detected.

4. Conclusion

The evolution of the microstructure after the heat treatment at the temperatures 300, 350 and 400 °C for 30 minutes followed by air cooling was investigated for NiTi (SE508 alloy) wires. Following conclusions were found by means of optical microscopy observation, XRD analysis and microhardness measurement:

1) The microstructure of all samples differed in morphology and was composed of needle/plate-like martensitic phase with different space distribution, thickness and length and remaining austenitic NiTi phase.
2) The heat treatment showed any influence on the occurrence or distribution of the $NiTi_2$ particles that remained aligned in parallel to the wire axis.
3) The microhardness of the samples in different heat treatment conditions differed very slightly and ranged from 373 to 406 HV0.1.
4) XRD analysis proved the existence of austenite NiTi and R-phase in all heat treatment conditions of the specimens. Only for the sample aged at 400 °C the monoclinic martensite was detected.

Acknowledgement

This work has been elaborated in the framework of project No. FV40306 "Development of new implants for correction of angular pediatric deformities in sterile design", supported by the Ministry of Industry and Trade of Czech Republic; project No. RRC/10/2017 "Support for Science and Research in the Moravia-Silesia Region 2017", financed from the budget of the Moravian-Silesian Region; SGS projects SP2019/128 and SP2019/43, financed by Ministry of Education, Youth and Sports of the Czech Republic.

References

[1] Bhagyaraj J, Ramaiah K V, Saikrishna, C N, Bhaumik S K and Gouthama 2013 Behavior and effect of Ti_2Ni phase during processing of NiTi shape memory alloy wire from cast ingot *J. Alloy Compd.* **581** 344-351

[2] Srivastava A K, Yang Z, Schryvers D and Van Humbeeck J 2008 Effect of annealing on cold-rolled Ni–Ti alloys *Mater. Sci. Eng.* A **481-482** 594-597

[3] Jiang F, Liu Y, Yang H, Li L and Zheng Y 2009 Effect of aging treatment on the deformation behaviour *Acta Mater.* **57** 4773–4781

[4] Zheng Y, Jiang F, Li L, Yang H and Liu Y 2008 Effect of ageing treatment on the transformation behaviour of Ti–50.9 at.% Ni alloy *Acta Mater.* **56** 736-745

[5] Jiang S, Zhang Y, Zhao Y, Liu S, Hu L and Zhao Ch 2015 Influence of Ni_4Ti_3 precipitates on phase transformation of NiTi shape memory alloy *Trans. Nonf. Metals Soc. China* **25** 4063-4071

[6] Adharapurapu R R and Vecchio K S 2007 Superelasticity in a new bioimplant material: Ni-rich 55NiTi Alloy *Exp. Mech.* **47** 365-371

[7] Elahinia M H, Hashemi M, Tabesh M and Bhaduri S B 2012 Manufacturing and processing of NiTi implants: A review. *Prog. Mater. Sci.* **57** 911-946.

[8] Losertová M, Štencek M, Matýsek D, Štefek O and Drápala J 2017 Microstructure evolution of heat treated NiTi alloys. *IOP Conf. Ser.: Mater. Sci. Eng.* **266** 012008

Influence of microstructure on fracture feature of Ti6Al4V alloy prepared by 3D printing

M Losertová[1], V Kubeš[1], J Drápala[1], K Dostálová[1], M A Sevostyanov[2], V Di Cocco[3] and F Iacoviello[3]

[1]VŠB – Technical University of Ostrava, Faculty of Materials Science and Technology, 17. listopadu 15/2172 , Ostrava – Poruba, Czech Republic
[2]Baikov Institute of Metallurgy and Material Sciences, 49 Leninski Prospects, Moscow, Russian Federation
[3]Università degli Studi Cassino e del Lazio Meridionale, Department of Mechanics, Structures, Environment and Territory, Viale dell'Universita, Cassino, Italy

mlosertova@vsb.cz

Abstract. In this work, the Ti6Al4V specimens produced by selective laser melting were mechanically tested in non-heat treated condition and fracture surface features were compared with this one of the material prepared using conventional casting and forging. The results of the fractography observation for both types of the samples were explained on the base of the microstructure analysis. The fracture surface of selective laser melted specimens showed more brittle feature that was in relation with the microstructure composed almost fully of martensite as opposed to conventionally prepared alloy with bimodal ($\alpha + \beta$) microstructure and more ductile character of fracture surfaces.

1. Introduction

New materials can be prepared by alternating both advanced technologies and alloying elements. Three biocompatible materials mostly used in traumatology or orthopedics are steels (ASTM F55 or F138), CoCr alloys (ASTM F75 or F90) and Ti6Al4V alloy (ASTM F136) [1-3]. Considering the increasing sensitivity of human body to nickel presence in materials, most of implants are produced now from Ti6Al4V alloy that possesses higher biocompatibility, higher corrosion resistance and favorable mechanical properties comparing with 316L stainless steel. Mechanical properties and strain distribution in relation to microstructure of Ti6Al4V resulting from mechanical and thermal treatment are important features influencing the reliability and safety of implants [3]. The Young modulus that is the critical parameter to avoid stress shielding in healing process, is in the case of compact Ti6Al4V implants produced by conventional method still high comparing modulus of human bone [1].

Biomechanical compatibility of Ti6Al4V can be improved using advanced producing methods, as for example selective laser melting (SLM). However, the material prepared by this method can display some defects, as pores, phase inhomogeneity and residual stresses [4-5] that can imply different mechanical and fracture behavior unlike conventional wrought Ti6Al4V products. Due to the vertical layer-by-layer build process during selective laser melting technology a highly directional microstructure in as-built condition comprises of fine acicular α' martensite that transforms to bimodal or lamellar ($\alpha + \beta$) equilibrium structure after heat treatment. Relieving residual stresses and modifying rapidly solidified microstructure provide a more desirable combination of strength and toughness.

The present work is focused on the evaluation of the fracture surfaces in relation with microstructure characteristics and mechanical properties that were determined for the as-forged and SLM as-built samples after tensile testing.

2. Experimental

Two sets of Ti6Al4V specimens were studied. First group of five tensile test samples (furthermore denoted as forged) was machined from traditionally forged rods (supplied by BIBUS METALS CZ). Second group represented five tensile test samples (furthermore denoted as SLM) that were prepared by selective laser melting (supplied by ProSpon spol. s r.o.) and machined before testing. The SLM samples were built in the vertical orientation (in the stress axis of tensile specimens) with specified parameters: laser power of 200 W, slice thickness of 30 μm, 5x5 mm islands rotated by 45 ° one from another, scanning speed of 800-900 mm/s.

Tensile testing of both sets was realized at the room temperature applying the force F = 17.3 kN and at the strain rate $\dot{\varepsilon} = 0.0025\ s^{-1}$ by GALDABINI, Sun 10 device. For the metallographic study, specimens were ground, polished and etched by means of the Kroll's solution (6 % HF, 8 % HNO_3, 86 % H_2O) for 10 - 60 seconds. The microstructure was studied using the optical microscope OLYMPUS GX51. The porosity was measured using an Image-Pro Plus software of Media Cybernetics. Fracture surfaces were observed by means of the PHILIPS SEM 505 microscope.

3. Results and discussion

The microstructure of Ti6Al4V forged samples in the as-received state is represented in figure 1. Transversal and longitudinal sections in figures 1a and 1b, respectively, show very fine grains of α phase with size under 5 μm (white phase) and (α + β) lamellar grains (dark grey phase) formed in prior β phase. Unlike cross section in the longitudinal direction the grains are elongated due to hot forging.

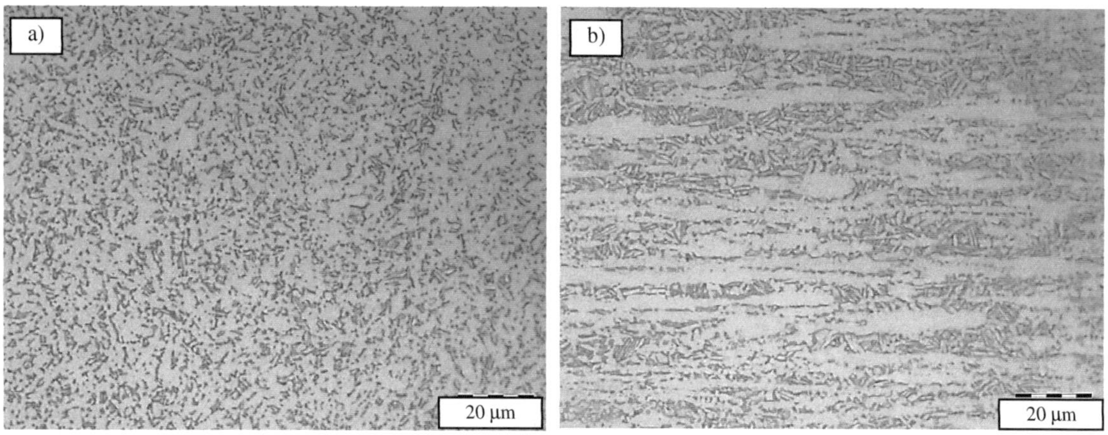

Figure 1. Microstructure of Ti6Al4V forged sample: a) bimodal morphology with fine (α + β) grains in the cross section and b) details of elongated α and β transformed grains in the longitudinal section.

The macrostructure of vertical cut of the SLM samples was characterized by patterns related with melted pools while in cross section the texture corresponding to scanning strategy (figure 2a) was observed. The microstructure of as-built specimens is formed of martensite needles in the prior β phase due to rapid cooling of melted pools, as seen in figure 2b. When heat treated the martensitic needles are transformed to α + β lamellar structure, as it was confirmed in [6]. The average value of the porosity measured in the as-built material (figure 2a) reached about 3 % whereas size pores varied from 2 to 159 μm.

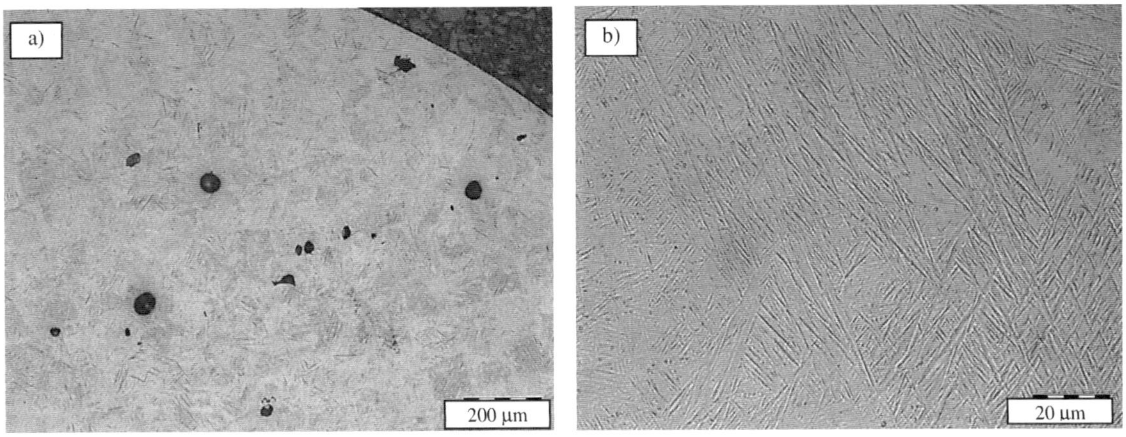

Figure 2. Microstructure of Ti6Al4V sample after SLM: a) cross section with pores, scanning strategy morphology and martensite needles; b) detail of acicular martensite structure.

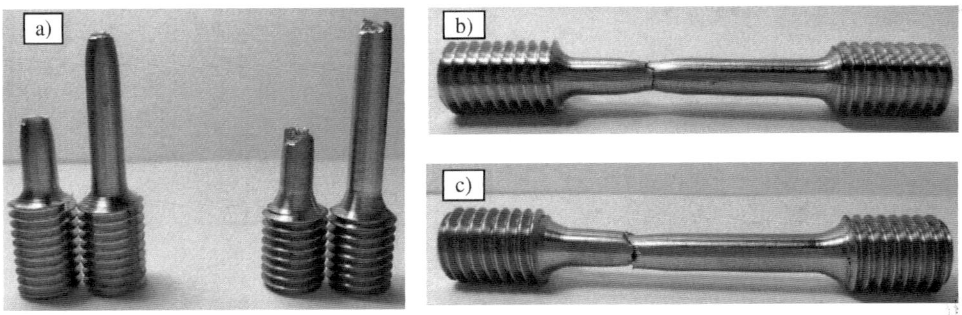

Figure 3. Macrographs: a) view of break after tensile testing of Ti6Al4V specimens (left – forged and right – SLM); b) and c) necking feature for forged (ε_f = 13.6 %) and SLM (ε_f = 7.2 %) samples, respectively.

Table 1. Mechanical properties of both sets of Ti6Al4V specimens in forged and SLM state.

Forged specimens	YS (MPa)	UTS (MPa)	ε_f (%)	*E* (GPa)
1	881	974	13.69	118
2	897	978	14.08	101
3	903	989	12.79	126
4	806	979	14.19	111
5	828	969	13.39	117
Average value	863±43	978±7	13.6±0.6	115±9
SLM pecimens				
1	1091	1271	7.23	93
2	1005	1260	7.82	95
3	924	1263	7.61	102
4	1052	1260	8.16	92
5	1033	1261	5.34	91
Average value	1021±63	1263±5	7.2±1.1	95±4

Note: YS – yield strength, UTS – ultimate tensile strength,
ε_f – strain to fracture; *E* – Young's modulus

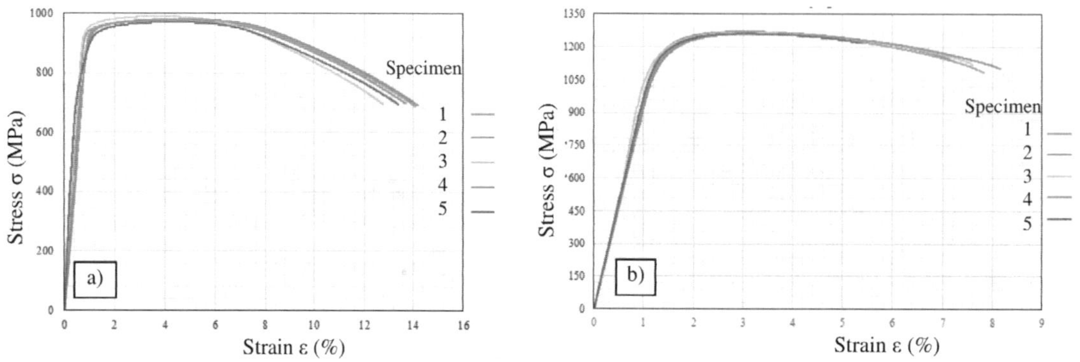

Figure 4. Tensile test diagrams of Ti6Al4V specimens: a) forged and b) prepared by SLM.

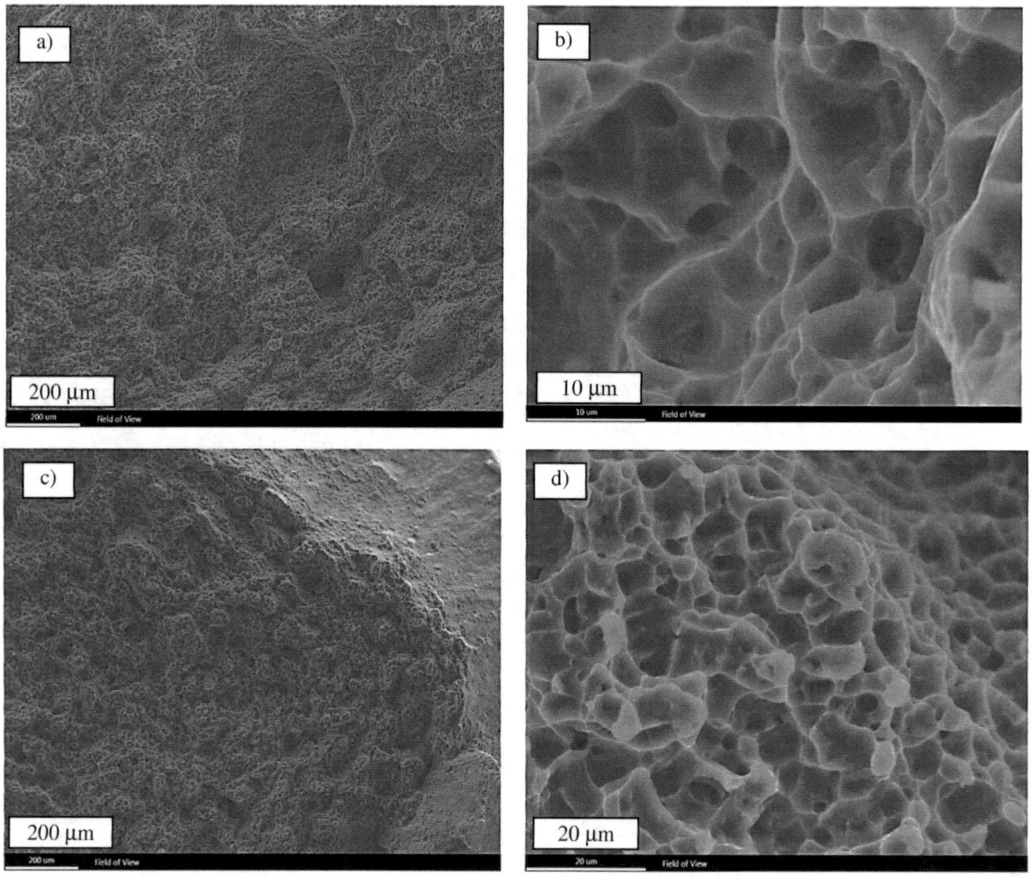

Figure 5. SEM fractographs of tensile tested Ti6Al4V specimens produced by forging: a) ductile failure in central part of fracture surface; b) detail of dimples and voids related with more brittle α phase; c) transition of ductile and shear fracture at the specimen edge; d) detailed deep dimples and shear facets at the transition area.

Character of the rupture differed for both sets, as seen in figure 3. The forged samples formed slight necking of the tensile samples (figure 3b). Due to the martensitic microstructure the SLM samples showed macroscopically more brittle feature of the damage than the forged samples.

The results of tensile mechanical properties of both specimen sets are summarized in table 1. Comparing the average mechanical values it is evident that martensite microstructure showed both higher yield strength (YS = 1021 MPa) and ultimate tensile stress (UTS = 1263 MPa). Unlike the bimodal (α + β) forged samples with YS = 863 MPa and UTS = 978 MPa, the martensitic microstructure of SLM material increased the average values of YS by 18 % and UTS by 29 %.

Figure 4 shows the typical tensile engineering stress - strain curves of both Ti6Al4V specimens sets tested to the fracture. As it can be seen, the deformation behavior of both materials with high strength characteristics differs under loading to the break. The bimodal (α + β) forged sample showed higher strain (ε_f = 13.6 %) comparing to the SLM sample with martensitic structure (ε_f = 7.2 %).

The SLM technology producing the porosity in the volume of the samples provided decreasing of the Young's modulus by 17 % (from 115 to 95 GPa - see table 1). The higher elasticity of the SLM material studied can be beneficial for medical applications but there is no evidence of the effect of the microstructure morphology and porosity volume on the fatigue and corrosion behavior.

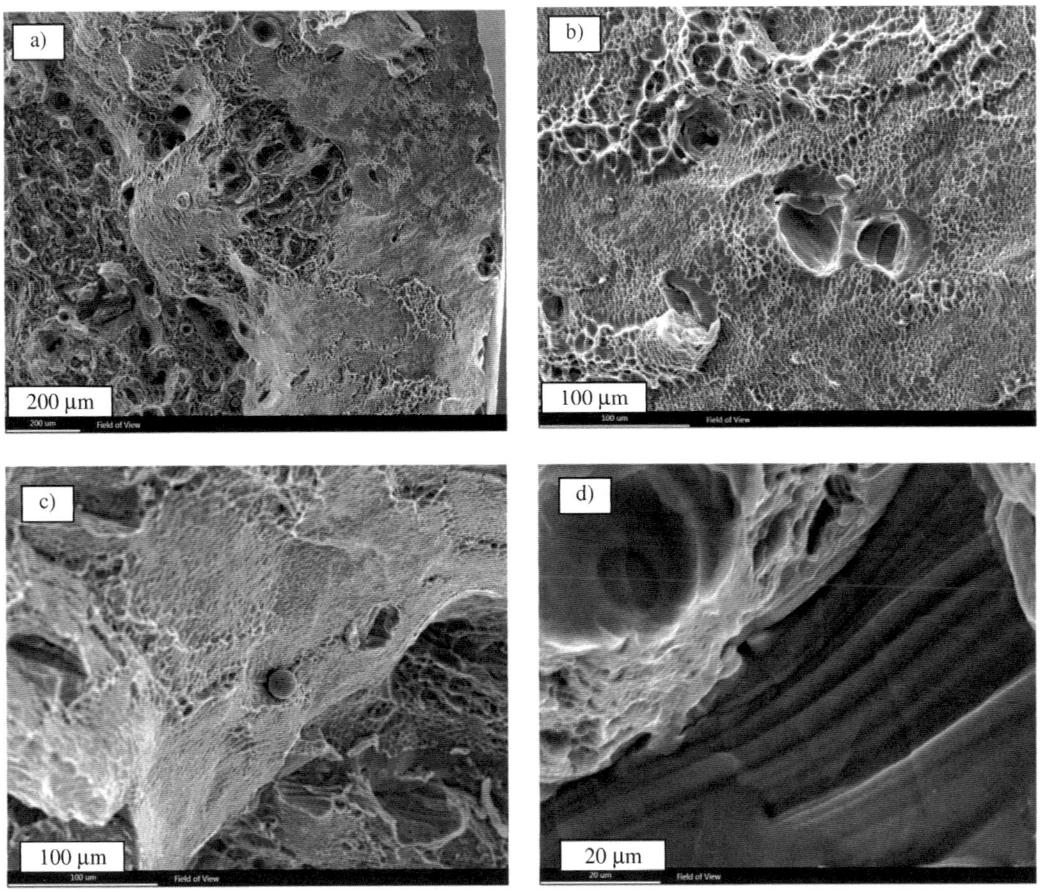

Figure 6. SEM fractographs of tensile tested Ti6Al4V specimens prepared by SLM method: a) shear fracture with pores at specimen edges; b) detail of very shallow dimples and pores on shear surface; c) detail of powder particle not remelted; d) detail of martensite laths and pore.

The evident differences in ragged fracture surfaces of tensile tested samples consisted in feature of dimples and shear facets (figures 5 and 6). Ductile failure with deep dimples is characteristic for the central part of the fracture surface in forged samples where existing holes are corresponding to more brittle α phase and transition area between ductile and shear fracture is observed at the specimen edge (figure 5a and 5b).

In some areas of SLM fractures, cavities with unmelted powder particles could be observed (figure 6b and 6c). Shallow dimples with cracked martensite needles rounded by ductile β phase in the fracture surfaces of the SLM samples (figure 6d) are corresponding to more brittle damage feature.

From the fractography study it is clear that SLM process-induced imperfections were the driving factor at the failure initiation on the microscopic scale. Moreover, the pores occurring in the specimen surfaces (figure 2a and 6a) served as notches localizing stresses at the loading. Generally, the Ti6Al4V alloy shows lower ductility comparing austenitic steels but the SLM process can contribute to further decreasing of the stress accommodation by plastic behavior due to providing the residual stresses, important porosity and metastable phases in rapidly solidified microstructure.

Conclusion

Based on the results of the comparative study of the tensile behavior, microstructure and fractography analysis of the Ti6Al4V specimens prepared of forged rods and by selective laser melting, it is possible to draw the following conclusions:

1) The as-received state of forged specimens was characterized by very fine bimodal microstructure composed of α and (α + β) grains due to previous mechanical treatment. Owing to this the mechanical properties reached 863 MPa and 978 MPa for YS and UTS, respectively. Due to the higher plasticity more ductile fracture feature with deep dimples were determined.
2) The SLM technology prepared specimens showed the microstructure consisted of fine martensite needles in prior β grains that led to the higher mechanical properties in YS and UTS (1021 and 1263 MPa, respectively) but the strain decreased by 47 %. According to the martensitic microstructure feature and porosity presence the damage at tensile tests was characterized by more brittle fracture feature.

Acknowledgement

This work has been elaborated in the framework of the Projects: No. FV40306 "Development of new implants for correction of angular pediatric deformities in sterile design", supported by the Ministry of Industry and Trade of the Czech Republic; RRC/10/2018 "Support for Science and Research in the Moravia-Silesia Region 2018", financed from the budget of the Moravian-Silesian Region; SGS projects SP2019/128 and SP2019/43, financed by Ministry of Education, Youth and Sports of the Czech Republic.

References

[1] Niinomi M 2008 Mechanical biocompatibilities of titanium alloys for biomedical applications *J. Mech. Behav. Biomed. Mater.* **1** 30-42
[2] Geetha M, Singh A-K and Asokamani R 2009 Ti based biomaterials, the ultimate choice for orthopaedic implants - A review *Prog. Mater. Sci.* **54** 397-425
[3] Craeghs T, Thijs L, Verhaeghe F, Kruth J-P and Van Humbeeck J 2010 A study of the microstructural evolution during selective laser melting of Ti–6Al–4V *Acta Mater.* **58** 3303-3312
[4] Enderle J-D, Bronzino J-D and Blanchard S-M 2005 *Introduction to Biomedical Engineering* Second ed. (Amsterdam: Elsevier Academic Press) 1253 p.
[5] Sun J, Yang Y and Wang D 2013 Parametric optimization of selective laser melting for forming Ti6Al4V samples by Taguchi method *Opt. Laser Technol.* **49** 118-124
[6] Losertová M and Kubeš V 2017 Microstructure and mechanical properties of selective laser melted Ti6Al4V alloy *IOP Conf. Ser.: Mater. Sci. Eng.* **266** 1 23

AUTHOR INDEX

Institute of Physics
Dirac House, Temple Back
Bristol BS1 6BE UK

ISSN: 1757-8981
ISBN 978-1-7138-2585-2

9 781713 825852

CURRAN ASSOCIATES INC.
proceedings.com